THE CONTRIBUTION OF SEED VARIETY TO CROP YIELD INCREASES IN CHINA

农作物品种增产贡献率测算研究

JIARONG QIAN

钱加荣◎著

中国农业出版社
CHINA AGRICULTURE PRESS
北　京
BEIJING

This study is funded by the National Social Science Fund of China (Grant No. 22BJY181).

PREFACE

Given the current agricultural conditions and constraints in resources and environment, further agricultural growth depending on significant increases in agricultural inputs is unlikely and potentially ineffective if achieved. Achieving food security is, therefore likely, to rely more heavily on the adoption and diffusion of new seed varieties in the future. Measuring the contribution of seed varieties to yield increases has thus become an essential concern of government to improve relevant national efforts and policies. However, estimating the contribution of seed varieties to actual yield increases nationwide remains a challenging issue for academics to tackle, despite this being an important reference for judging the ongoing development of breeding technologies, and being of great significance in developing and improving seed industry revitalization policies.

This study develops a method to estimate the contribution by splitting the study period in terms of replacement in the main diffused varieties and specifying a yield response model that incorporates a series of dummy variables to capture net increases attributed to variety diffusion in each sub-period. Using this method, the contributions of seed varieties for six main crops in China have thus been estimated.

In a departure from most previous studies, this method aims to estimate the overall actual contribution of such varieties to yield gains, thus attributing improvements from seed varieties to both breeding and diffusion. Besides, official statistical yield data at the farm level are used for estimation, allowing the estimated contribution more accurately represents the actual impact of seed varieties on crop yields, addressing the existing problem of a gap between trial and actual yields encountered by most previous studies.

This method has several advantages, such as being concise, easy to use, and flexible. Furthermore, it can distinguish seed varieties' historical contributions and can be used in tracking assessments. Therefore, it promises to be a good method for applying to the practical measurement of progress in seed breeding technologies and agricultural development.

Jiarong Qian
July 26,2023

ABSTRACT

Sustainable development in agriculture relies on continuing progress in variety breeding and adoption. Measuring the contribution made by new seed varieties to yield increases nationwide is thus essential; however, this process remains challenging for academics. This study thus develops a method that enables the contribution to be estimated based on splitting the study period in terms of replacement in the main diffused varieties and the specification of a yield response model that incorporates a series of dummy variables to capture net increases attributed to varieties diffused in each sub-period, with the Arellano-Bond estimation method then applied to overcome the endogeneity of the model and to thus obtain more precise estimates of contribution. Using this method, the respective contributions of varieties to yield increases for rice, wheat, corn, soybean, cotton, and rapeseed in China during 1980 and 2019 were estimated.

According to the empirical results from the proposed method, the absolute contributions of current varieties over the base period were estimated to be 1,165.3 kg/hm^2, 486.7 kg/hm^2, 1,026.8 kg/hm^2, 329.4 kg/hm^2, 215 kg/hm^2, and 416.1 kg/hm^2 for rice, wheat, corn, soybean, cotton, and rapeseed, respectively, giving respective contribution rates of 48.09%, 33.78%, 48.52%, 47.35%, 37.15%, and 44.32%. The results illustrate that the absolute contribution to crop yield increases has maintained steady growth, with the total increases reaching 107.2%, 106.5%, 152.3%, 357.5%, 530.5%, and 300.1% for the six crops, respectively. However, contribution rates for the main grain crops have seen a declining trend; the contribution rates for rice and corn were 70% and 60% at the end of 1980s and in the mid-1990s, while current contribution rates have decreased to 48.1% and 48.6%, decreases of 25.3 percentage points and 10.5 percentage points from the recorded high levels, respectively. This decline in contribution rates may be attributable to the extensive usage of input factors, which generates a large proportion of such yield increases; the current usage of chemical fertilizers, pesticides, and other materials significantly exceeds the input levels for the 1980s and 1990s, and the extensive input of such materials significantly enhances crop yield levels and reduces the contribution share of seed varieties.

It is suggested that the administrative departments increase support for scientific and technological innovation in the seed industry alongside building a modern diffusion system for varieties, vigorously supporting the breeding of environment-friendly varieties, formulating matched support facilities for excellent varieties and suitable planting patterns, and giving full play to an investigation into the key supporting role of varieties in ensuring national food security and realizing high-quality agricultural development.

CONTENTS

PREFACE

ABSTRACT

CHAPTER 1   Introduction ··· 1

   1.1   Background ·· 1
   1.2   Literature review ·· 2
   1.3   Methodology ·· 10
   1.4   Organization ·· 21

CHAPTER 2   The contribution of seed variety to rice yield increases ·· 23

   2.1   Rice yield changes in China ··· 23
   2.2   Evolution in rice seed varieties in China ····································· 24
   2.3   Specification for rice yield model ··· 25
   2.4   Model estimation ··· 26
   2.5   Empirical results ·· 28
   2.6   Summary ··· 34

CHAPTER 3   The contribution of seed variety to wheat yield increases ·· 36

   3.1   Wheat yield changes in China ·· 36
   3.2   Evolution of wheat seed varieties in China ·································· 38
   3.3   Specification for the wheat yield model ······································ 39

3. 4　Model estimation　39
3. 5　Empirical results　41
3. 6　Summary　46

CHAPTER 4　The contribution of seed variety to corn yield increases　49

4. 1　Corn yield changes in China　49
4. 2　Evolution in corn seed varieties in China　51
4. 3　Specification for the corn yield model　52
4. 4　Model estimation　52
4. 5　Empirical results　55
4. 6　Summary　59

CHAPTER 5　The contribution of seed variety to soybean yield increases　61

5. 1　Soybean yield variation in China　61
5. 2　Evolution in soybean seed varieties in China　63
5. 3　Specification for soybean yield model　64
5. 4　Model estimation　64
5. 5　Empirical results　67
5. 6　Summary　71

CHAPTER 6　The contribution of seed variety to cotton yield increases　73

6. 1　Cotton yield variation in China　73
6. 2　Evolution in cotton seed varieties in China　74
6. 3　Specification for the cotton yield model　75
6. 4　Model estimation　76
6. 5　Empirical results　78
6. 6　Summary　83

CONTENTS

CHAPTER 7 The contribution of seed variety to rapeseed yield increases 85

7.1 Rapeseed yield changes in China 85
7.2 Evolution in rapeseed seed varieties in China 86
7.3 Specification for rapeseed yield model 87
7.4 Model estimation 88
7.5 Empirical results 90
7.6 Summary 95

CHAPTER 8 Conclusion and policy recommendations 97

8.1 Conclusion 97
8.2 Policy recommendations 103
8.3 Suggestions for further research 107

REFERENCES 108

CHAPTER 1   Introduction

1.1   Background

Seeds are the basis of agriculture, and play a critically important role in food security strategies that may affect the future and destiny of a country. The development of seed industry has recently been raised to an unprecedented new height in China. In 2020, the Central Economic Work Conference proposed to overcome "the 'bottleneck' technology of seed resources to win the battle of the seed industry". Sustainable development in agriculture now relies on ongoing progress in seed variety breeding and adoption. However, as against a background of increasingly tight constraints on agricultural resources and environmental pollutants, it is now impossible to increase grain total production through the use of extensive input factors as in the past. The optimal way to achieve progress is thus to develop and adopt new excellent varieties and to give full play to the yielding potentials of varieties. Therefore, it is important to be able to scientifically estimate the contribution of novel seed varieties to crop yield increases to incorporate this information into agricultural and technological policy design. However, measuring seed varieties' contribution to overall yield increases nationwide is challenging because of a lack of suitable variables that reflect the seed varieties' potential in enhancing yield.

The contribution rate of varieties to yield increase is an important indicator for measuring the development status of the seed industry. In the new stage of development, taking into account the constraints of resources and environment, it is particularly important to study the contribution of seed varieties to crop yield increases. Developing a scientific measurement for the contribution of varieties thus has important implications for designing better policies to support seed industry development. However, this remains challenging for academics, and relevant research still needs to be conducted. This study thus aims to present a method allowing the overall contribution of seed varieties to yield increase to be accurately estimated to provide empirical indicators for measuring the development level of the seed industry, which is of great practical significance for improving breeding science and technology, supporting the progress of agricultural technology, ensuring national food security, and achieving high-quality development of agriculture.

1.2　Literature review

1.2.1　Studies using multiple regression

Earlier studies often estimated the impact of seed varieties or genetic improvements on crop yields by employing multiple regression analysis under the framework of a production function. Brennan (1984) constructed three measures—"the index of varietal newness", the proportion of the area planted to recently released varieties, and the "index of varietal improvement" —to represent the yield potentials of varieties, and investigated the contribution of seed varieties to wheat yield increases by incorporating the three measures into a production function. Feyerherm et al. (1984) investigated the genetic contribution of seed varieties to wheat yield increases by employing a differential yielding ability (DYA) value established by computing the mean difference in yields between the given variety and a primary check variety over a set of years and locations within a geographical region of mutual adaptability.

Hu et al. (2002) proposed a new index of genetic uniformity instead of the coefficient of parentage (COP) to reflect genetic diversity, and constructed a common production function to estimate the effects of genetic diversity on wheat yield; the model included the index of genetic uniformity, as well as considering technological progress and various input factors and environmental factors. The empirical results, based on data from wheat varieties approved between 1982 and 1997 across 15 major production areas in China, revealed a negative relationship between genetic uniformity and wheat yield, implying that wheat production should avoid adopting the varieties with the same genetic relationships to sustain stable wheat yields. Duvick (2005) noted that an Iowa-adapted time series of hybrids representing the period from 1930 to 2001 showed a linear gain for grain yield of 77 kg/hm^2 per year, based on a regression analysis using trial data. Duvick et al. (2005) further provided an estimate of a 51% contribution from genetics when trial yields were adjusted to the equivalent of average on-farm yields for Iowa from 1930 to 2001.

Si and Li (2018) calculated the improved varieties rate (IVR) and varieties renewal rate (VR) to reflect the popularization of improved soybean varieties based on the provincial sowing areas for soybean varieties from 1985 to 2014. They then incorporated the IVR and VR into a production function that considers quadratic variables and time trends. The estimated results suggested that the IVR and VR had a significant positive impact on soybean yield, such that a 1% increase in each of these indexes will result in a 0.021% and 0.018% increase in soybean yields, respectively.

1.2.2 Studies using the Just-Pope production approach

Just and Pope's econometric framework relaxes the second moment production function restrictions, thus providing a method for estimating the marginal risk effects of explanatory variables. The Just-Pope production function has thus been applied to estimate the contribution of seed varieties, estimating the stochastic production functions allowed by varieties. Traxler et al. (1995) constructed a Just-Pope production function to explore the effects of genetic improvement on the mean and variance of wheat yields using trial data from a variety x nitrogen experiment conducted by the International Maize and Wheat Improvement Center. The model considered ten wheat varieties, released in Mexico from 1950 to 1985, and set five levels of nitrogen fertilization. The mean and variance of wheat yields were specified as functions of nitrogen and technology, while the log of the variance of yield was specified as a quadratic functional form. The study concluded that the increase in the mean wheat yield slowed during the post-green revolution era, but that improvement in yield stability remained relatively rapid. Overall, steady progress in producing better varieties was reported.

Smale et al. (1998) employed a Just-Pope approach to estimate the impact of genetic resources on the mean and variance of wheat yields by incorporating several indicators of genetic diversity into the function to capture the effect of genetic diversity. The mean and the variance functions were estimated using a three-stage feasible generalized least squares (FGLS) procedure applied to cross-sectional time series of wheat production data for 29 districts of Punjab, Pakistan, from 1980 to 1986. The estimation results revealed that higher area concentration among varieties is associated with higher mean yields in irrigated areas, while genealogical variables are associated positively with mean yield and negatively with yield variance in rainfed districts.

Wang et al. (2001) measured the genetic diversity of soybean varieties released between 1980 and 1997 using the coefficient of parentage (COP) and WCOP, thus examining the relationship between genetic diversity and soybean yield stability by employing a Just-Pope production function that incorporated genetic diversity as an explanatory variable to capture the effect of genetic diversity on both mean yield and yield stability for soybeans in China. The estimation results indicated that the relationship between yield and genetic diversity follows a parabola pattern, with the yield being most conductive at a certain coefficient level (COP = 0.122, WCOP=0.096). Furthermore, the significant positive relationship between yield variance and genetic diversity suggested that genetic diversity was required for yield stability. The most effective methods for ensuring further improvement and sustainable development of soybean production were thus identified as timely and effective protection and reasonable utilization of soybean genetic diversity.

1.2.3 Studies using the Analytic Hierarchy Process method

Several studies have also employed the Analytic Hierarchy Process (AHP) method to assess the contribution of seed in terms of agricultural technological change; AHP is an effective tool for dealing with complex assessments by reducing complex factors to a series of pairwise comparisons and then synthesizing the results based on relative pairwise evaluations (both qualitative and quantitative) made by decision-makers (Saaty, 1980). Saaty and He (1995) used AHP to analyze the contributory share of each factor affecting agricultural technological change in Youzhong County. The results suggested that seed variety accounted for the largest share of total technological progress at 25.6%. Zhao and Zhang (2005) employed AHP to evaluate the contribution of each specific factor affecting agricultural technological change in China, concluding that seed variety made the largest contribution among all technological aspects at 29.6% of the total contribution. Although a general assessment of seed contribution can be made using the AHP method, AHP requires analysts to set weighted values for each factor; such values depend on the subjective judgments of the analysts, who may reach different conclusions. The AHP method further depends on pairwise relative evaluations made by decision-makers, therefore, cannot produce an objective result. Nevertheless, it remains an effective tool for determining the most important factor based on an informed ranking of all known influence factors.

1.2.4 Studies estimating yield potentials

Evans and Fischer (1999) analyzed the definition, measurement, and significance of yield potential, taking yield potential to refer to the yield of a variety when grown in an environment to which it is well adapted, under circumstances where nutrients and water are not restricted and where pests, diseases, weeds, collapses, and other stresses are effectively controlled. Giunta et al. (2007) determined the impact of changes in yield potential due to durum wheat breeding in Italy on future breeding objectives for durum wheat improvement in the Mediterranean environment by comparing the yields of 20 durum wheat varieties (before 1950), intermediate varieties (1950 to 1973), and modern varieties (1974 to 2000) in an irrigated field trial with two sowing periods and two N applications. It was found that the average yield of the intermediate varieties was 39% higher than those of the old varieties but 18% lower than those of the modern varieties.

Neumann (2010) calculated global datasets of maximum attainable grain yields, yield gaps, and efficiencies of grain production at a spatial resolution by employing a stochastic frontier production function integrating biophysical and land management-related factors. The results showed that some regions produce grain close to the estimated frontier yields while others demonstrate a large yield gap of up to 7.5 tons/hm^2 for wheat, 8.4 tons/hm^2 for corn, and 6.4 tons/hm^2 for rice, indicating the potential to increase actual grain yields in those regions. Moreover, the individual contributions of regional determinants of efficiencies, such as irrigation, market accessibility, market influence, agricultural population, agricultural labor, and field slope, varied strongly between world regions and grain types.

Wang et al. (2011) simulated the yield potentials of summer corn in the Beijing-Tianjin-Hebei (BTH) region by employing the WOFOST model, which considered daily weather data from 1968 to 2007. Their results suggested that the summer maize potential yield was 6,854 to 8,789 kg/hm^2, roughly increasing from the west and south to the northeast. The water-limited yield was 6,434 to 8,741 kg/hm^2, again increasing from south to northeast. According to the statistical data at the county level, the actual yield of summer maize thus still has a 30% gap for the potential yield.

Patrignani et al. (2014) used the data for grain yields in 19 counties in Oklahoma to estimate the yield gap by using three different yield calculation methods and fitting the grain yield and rainfall in the growing season into the frontier yield functions. The results found that the experimental yield of Winter Wheat Varieties in Oklahoma under rainfed conditions could be as high as 6.59 kg/hm^2, while the maximum yield under southern plain irrigation was 7.69 kg/hm^2; the current yield, however, was only 2.0 kg/hm^2. The gaps between the average unit yield of the state and the achievable yield and the potential yield of water restriction were thus 0.7 and 4.7 kg/hm^2, respectively. Currently, Oklahoma's grain yield may reach 74% of the attainable yield, but only 30% of the water-limited yield of the state. The current state-level production is 77% of attainable production, which is 31% of the water-limited production of the state.

1.2.5 Summary of literature

To investigate the contribution of seed varieties to crop yield increases, many studies have focused on exploring the contribution of genetic improvements. However, national yield increases are only partially dependent on genetic improvements in new seed varieties; the correct diffusion of improved seed varieties is also a very important aspect of enhancing crop yields nationwide. One way to enhance yield level is to improve seed varieties, while another is to diffuse such improved varieties over large areas. Most existing studies did not refer to the contribution of variety diffusion to yield increases. The overall impact of seed varieties in terms of genetic improvement in seed varieties and diffusion of these improved varieties on nationwide yield level has not been well documented in previous studies.

Several studies have employed the AHP method to assess the contribution of seed in terms of agricultural technological change; this is an effective tool for dealing with complex assessments by reducing the complex factors to a series of pairwise comparisons and then synthesizing the results based on the relative pairwise evaluations (both qualitative and quantitative) made by decision-makers. However, although a general assessment of seed contribution can be made using the AHP method, AHP requires analysts to set weighted values for each factor. Those values depend on the subjective judgments of the analysts, who may reach different conclusions. It is thus impossible for this method to generate objective results.

In addition, most previous studies used a variety of performance trial data to investigate the seed contribution, so the estimations create problems in terms of extrapolating from performance in nurseries to those seen in actual growing conditions (Feyerherm et al., 1984). Results based on such trials are likely to overestimate the yield improvement of using the new varieties on the actual farm level (Brennan, 1984). Measuring the actual variety contribution in nationwide or regional planting should consider the effects of weather, disease, and varying production conditions over large areas, which thus remains highly uncertain. Nevertheless, estimating this actual contribution is very important for future planning purposes, and developing better ways to estimate this overall contribution also contributes to the literature.

This study aims to develop a method from the perspective of agricultural economics to investigate the contribution of new seed varieties to nationwide crop yield increases. To facilitate this, the term contribution of crop varieties as used within this study will refer to the contributions made by genetic improvement in crop varieties and the diffusion of the improved varieties. This study contributes practical and heuristic significance in measuring the developmental status of both seed breeding technologies and agriculture.

1.3 Methodology

1.3.1 Crop variety diffusion

Normally, new varieties are developed and diffused into

agricultural production regularly, and their diffusion may last for a few years. These new seed varieties' yielding ability can therefore only be fully demonstrated during their entire diffusion period. Estimating the contribution of seed varieties throughout the entire diffusion period is more scientific and reasonable than studying only one instance. To assess the contribution of seed varieties to yield increases, the study period can be divided into several diffusion periods, with regards to the replacement of the major varieties and estimations made of the overall contribution in each diffusion period.

Figure 1-1 illustrates the evolution of diffusion for crop seed varieties. If there is no obvious change in major seed varieties from t_0 to t_1, then we can set period $t_0 - t_1$ as Period 1 (the base period). If new seed varieties are extensively planted in period $t_1 - t_2$, this period can be identified as Period 2. The remaining periods can be deduced by analogy until Period n.

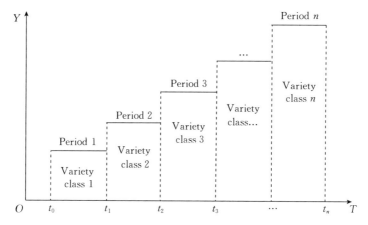

Figure 1-1 Seed diffusion periods for various crop varieties

Figure 1-2 presents the yield curve for each period. Supposing the new diffused seed varieties increase crop yield by an average of h. The yield curve will move upward by h, So the intercept will increase by h. Any decreases will decrease the intercept. Therefore, the h represents the contributory proportion of new seed varieties to crop yield increases, the key to measuring new seed varieties' contribution is to estimate the h.

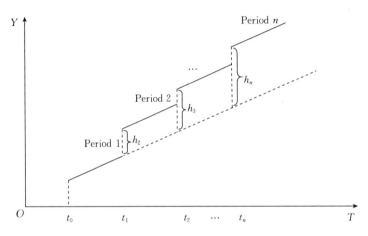

Figure 1-2 Yield curves in each diffusion period for varieties

1.3.2 Inference for yield model

Although previous studies have developed various yield models to predict crop yields (Feyerherm et al., 1981; Bell et al., 1994; Zere et al., 2005; Mirschel et al., 2014), these have several limitations. Taking into account that: (1) sufficient data for constructing a production function for yield is generally unavailable, (2) crop yield is significantly sensitive to crop producer price which is the average price or unit value received by farmers at the farm gate (Qian et al., 2012; Qian et al., 2015), and (3) a supply response model that contains crop producer prices is powerful in explaining changes in supply variables (area, yield, and production), the crop yield model used in this study is specified as a yield response equation.

The response equation is derived from the adaptive expectation theory. This theory links farmers' behaviors to their expected prices and posits that farmers react not to the previous year's price but rather to the price they anticipate in the current year, and that this expected price depends only to a limited extent on the previous year's price (Nerlove, 1956; Nerlove, 1960). Formally, the adaptive expectation is written as

$$P_t^e = P_{t-1}^e + \lambda (P_{t-1} - P_{t-1}^e) \qquad (1\text{-}1)$$

where P_t^e is the expected price in year t; P_{t-1}^e represents the expected price in year $t-1$; P_{t-1} represents the actual price in the previous year; and λ is the coefficient of expectation with $0 < \lambda \leqslant 1$, which reflects how much information producers retain in their current year's expectations from outcomes observed in the previous year. Farmers will revise their expected price in terms of the difference between the previous year's actual price and the expected price, rather than simply using the previous year's price as the current year's expected price (Vitale et al., 2009).

The basic yield equation can be written as

$$Y_t = \alpha_0 + \alpha_1 P_t^e + \mu_t \qquad (1\text{-}2)$$

where Y represents crop yield; P_t^e is the expected producer price for the crop; μ is a random residual term, t is the year; and α_0 and α_1 are coefficients to be estimated. Unfortunately, the expected price cannot be observed, and thus these coefficients cannot be estimated. To solve this problem, we include Equation 1-3, the basic yield equation in the previous year.

$$Y_{t-1} = \alpha_0 + \alpha_1 P^e_{t-1} + \mu_{t-1} \qquad (1\text{-}3)$$

Combining the adaptive expectation Equation 1-1 and the yield Equations 1-2 and 1-3 leads to a basic yield response model of

$$Y_t = \alpha + \beta_0 Y_{t-1} + \beta_1 P_{t-1} + v_t \qquad (1\text{-}4)$$

where α turns out to be equal to $\alpha_0 \lambda$, β_0 equals $1-\lambda$, β_1 equals $\alpha_1 \lambda$, and v is a random residual that differs from μ. Most importantly, all variables are observed; therefore, parameters can be estimated using observed data. This model is so-called Nerlove model, which has been widely applied to estimate this dynamic process in crop production (French et al., 1971; Askari et al., 1977; Gafar, 1987; Froster et al., 1995; Wang et al., 1998; Mushtaq et al., 2002; Qian et al., 2012; Miao et al., 2016).

To capture the contribution of a different class of seed varieties to increases in crop yields, a series of dummy variables representing various diffusion periods are incorporated into the basic yield response model in combination with the earlier analysis on a division of various variety diffusion periods, and parameters on these dummy variables capture the net average increases over the base period attributed to the adoption of the improved varieties in each diffusion period. The resulting extended yield response model is shown as

$$Y_t = \alpha + \beta_0 Y_{t-1} + \beta_1 P_{t-1} + \beta D + v_t \qquad (1\text{-}5)$$

where D indicates a series of dummy variables distinguishing diffusion periods for various seed varieties. Here β represents the net increase caused by new seed variety diffusion. If the adoption and diffusion of new seed varieties significantly promote crop yield, the overall average yield during the diffusion period will rise, and β, the coefficient on dummy variables is expected to take a positive sign; otherwise, a negative sign is expected.

1.3.3 Estimation strategy

The yield equation is a dynamic panel model containing a lagged dependent variable. This may lead to an endogeneity problem, as the error term may be correlated to the lagged dependent variable. When estimating dynamic panel models, the Ordinary Least Squares (OLS) estimation method does not work due to endogeneity. The Arellano-Bond estimator (Arellano et al., 1991) thus instead exploits the lagged endogenous variables as instruments without sourcing the instrument variables from outside data, producing unbiased, consistent, and efficient estimators and providing additional flexibility in practical implementation. The Arellano-Bond estimation method was thus used to estimate the coefficients to overcome the risk of endogeneity and improve the estimates of the dynamic panel models.

The Arellano-Bond method is based on the Generalized Method of Moments (GMM), which uses lagged endogenous regressors as instrument variables. The Arellano-Bond procedure takes a first-difference of the levels' equation to remove the fixed effects of the cross section, and then uses the lagged endogenous regressors as instruments to form moment conditions. This causes the endogenous variables to be pre-determined and not correlated with the error term.

The general dynamic panel model is expressed as

$$y_{it} = \alpha y_{it-1} + \beta x_{it} + u_{it} \qquad (1\text{-}6)$$

$$\Delta u_{it} = \Delta v_{it} + \Delta e_{it} \qquad (1\text{-}7)$$

where y_{it} indicates the independent variable of cross section i in year t; x is a vector of explanatory variables; and u is the error term, which consists of fixed cross section effect v and the residuals. The Arellano-Bond method uses the first-differences approach to transform Equation 1-6 into Equation 1-8 by transforming the regressors using first differencing so that the fixed cross-specific effect contained in the error term is removed and the first-differenced lagged dependent variable is also instrumented with its past levels.

$$\Delta y_{it} = \alpha \Delta y_{it-1} + \beta \Delta x_{it} + \Delta u_{it} \qquad (1\text{-}8)$$

$$\Delta u_{it} = \Delta v_{it} + \Delta e_{it} \qquad (1\text{-}9)$$

1.3.4 Calculation for varietal contribution

The mathematical proof process of estimating seed contribution to yield increase are provided. To simplify the equation, the crop yield equation is defined as

$$Y = C + \alpha X + \beta D \qquad (1\text{-}10)$$

where Y indicates crop yield; X represents the main variables influencing crop yield; D represents a dummy variable reflecting changes in seed diffusion; and C is the intercept. Suppose the first diffusion stage, t, is from year 0 to k, and the nth stage when the major diffused seed varieties are changed is from year i to $i+n$. The individual yield equations for each year in both periods are therefore as follows

$$Y_0=C+\alpha X_0 \quad Y_i=C+\alpha X_i+\beta$$
$$Y_1=C+\alpha X_1 \quad Y_{i+1}=C+\alpha X_{i+1}+\beta$$
$$Y_2=C+\alpha X_2, \quad Y_{i+2}=C+\alpha X_{i+2}+\beta \quad (1\text{-}11)$$
$$\vdots \qquad\qquad \vdots$$
$$Y_k=C+\alpha X_k \quad Y_{i+n}=C+\alpha X_{i+n}+\beta$$

By adding these two groups of equations, we obtain

$$\sum_{t=0}^{k}Y_t = kC + \alpha \sum_{t=0}^{k} X_t \quad (1\text{-}12)$$

$$\sum_{t=i+1}^{i+n}Y_t = nC + \alpha \sum_{t=i+1}^{i+n} X_t + n\beta \quad (1\text{-}13)$$

then, the average yield functions for the 1 and n periods are obtained by dividing by k and n for both sides of Equation 1-12 and Equation 1-13, respectively. These are expressed as follows

$$\frac{1}{k}\sum_{t=0}^{k}Y_t = C + \frac{\alpha}{k}\sum_{t=0}^{k} X_t \quad (1\text{-}14)$$

$$\frac{1}{n}\sum_{t=i+1}^{i+n}Y_t = C + \frac{\alpha}{n}\sum_{t=i+1}^{i+n} X_t + \beta \quad (1\text{-}15)$$

Subtracting Equation 1-14 from Equation 1-15, the increase in average yield in period n over period 1 is expressed as

$$\Delta Y = \frac{1}{n}\sum_{t=i+1}^{i+n}Y_t - \frac{1}{k}\sum_{t=0}^{k}Y_t = \alpha\left(\frac{1}{n}\sum_{t=i+1}^{i+n} X_t - \frac{1}{k}\sum_{t=0}^{t=k} X_t\right) + \beta$$

$$(1\text{-}16)$$

where ΔY indicates the increase in average yield from Period n to Period 1; k, n, Y, and X are observed variables; α and β are parameters that can be estimated, and β is the increase caused by new seed varieties. Finally, the contribution rate of new seed varieties can be calculated by

$$CR = \frac{\beta}{\Delta Y} \times 100\% \quad (1\text{-}17)$$

In addition, this method can be extended to make a comparison between any two diffusion periods; the contribution formula is

$$CR = \frac{\Delta \beta}{\Delta Y} \times 100\% \qquad (1\text{-}18)$$

The mathematical proof process for this Formula 1-18 is similar to that demonstrated for the previous Formula 1-17.

1.3.5 Advantages of the method

Researches estimating the contribution of seed variety to yield increase are few. The proposed method focuses on the impact of various changes on crop yields while controlling for the influences on the yield level from other factors through using a one-year lagged crop yield, which ensures that impacts of other factors such as chemicals, labor input, cultivation conditions, field management, agricultural policy, and weather, are not mistaken for various impacts. It can thus be applied to the practical estimation of variety contribution for various crops or livestock. More generally, the estimation method has the following advantages:

(1) High stability. The model adopts the form of a supply response equation, which takes a simple structure, making it easy to estimate model parameters, and offering strong robustness to the estimation results.

(2) Strong explanation power. The model well controls factors affecting crop yield, and can thus accurately explain variations in yield. This offers strong interpretation and prediction ability.

(3) Easy access to estimation data. The data involved in estimation include crop yield, crop producer price, and the

diffusion of leading varieties. These data can be easily obtained from official statistical yearbooks or reports, allowing this method to be readily utilized in practical applications.

(4) Available for tracking assessment. The subsequent contribution indicator can be calculated annually for reporting purposes by simply updating the annual data, satisfying governmental demands for tracked assessment of the progress of breeding programs and the work of variety diffusion. In addition, this method can distinguish the historical contributions of seed varieties by comparing the contribution between any two diffusion periods.

Overall, the proposed method achieves the goal of confirming the contribution of varieties to yield increases, supporting the growth of related policies for agricultural development, particularly within the seed industry. Therefore, it could be designed to support regular tracking assessments, making this a useful tool to be applied to effectively assess the role of seed varieties in the strategy of food security.

1.3.6 Discussion

Yield gains nationwide depend on genetic improvement in varieties, but they are also affected by the diffusion of those improved varieties. These connected factors mean that only the overall contribution of varieties, including genetic improvement and diffusion, can be used to reflect the progress in seed development and spread in a country. While most previous studies used varieties' performance trial data to investigate the contribution of genetic improvement in varieties, this creates problems in terms of extrapolating from nurseries to actual growing conditions. Thus, results based on such trials are likely to overestimate the yield improvement on farms from using the new varieties (Brennan, 1984). The impact of such tested varieties, taking into account the effects of weather, disease, and varying production conditions in large areas, remains to be determined. Nevertheless, its estimation is very important for planning purposes, and finding better ways to measure the overall actual contribution makes significant sense regarding governmental and political decision-making.

In a departure from most previous studies, this study developed a method aiming to estimate the overall contribution, rather than only the genetic contribution, to yield gains that can be attributed to new seeds from improvements in both breeding and diffusion as estimated by capturing changes in the intercept of the yield curve. Other factors such as weather variability, disease, input usage, and field management also affect crop yields, and controlling for these factors is problematic for those studies using trial data (Bell et al., 1995). In this instance, these are comprehensively represented by a technical process employing a proxy variable of lagged yield (Wooldridge, 2013; Wang et al., 2007). Further, farm field yield data is used for estimation; thus, the estimated contribution represents the impact of new varieties. The coefficients represent intercept movement. Theoretically, the coefficient could be negative, if the estimated value took a negative sign, indicating that the yield curve was reducing, and that new seed varieties were failing to enhance crop yields. In such a case, the government should develop new varieties with high yield ability or try to expand the diffusion areas of the new varieties to enhance the seed varieties' contribution.

The estimation uses a relatively large sample size of provincial panel data, which improves the precision and stability of the coefficients, and thus the reliability of the results. Increases in crop yields due to new varieties diffusion had been captured in each diffusion period. Also, the historical contributions of seed varieties can be distinguished. The suggested method involves only a few variables, making it uncomplicated, and the data can be easily collected from yearbooks or official reports. This method can be easily utilized in practical applications based on these facts. In addition, the contribution indicator thus generated can be reported annually by updating the annual data, satisfying the governmental demand for annual assessment of the progress of the breeding program and the work of new variety diffusion. Corresponding policies for agricultural development, particularly for the seed sector, could therefore be designed concerning these tracking assessments.

1.4 Organization

This study contains two main sections. Chapter 1 is the first section, which states the background, objectives, along with a literature review. It then presents the proposed method for estimating the contribution of seed varieties to crop yield increases. The second section includes Chapters 2 to 8. Chapters 2 to 7 discuss the calculated seed variety contributions for rice, wheat, corn, soybean, cotton, and rapeseed, respectively, based on the method presented in section one; while Chapter 8 presents the conclusion and policy recommendations.

CHAPTER 2　The contribution of seed variety to rice yield increases

2.1　Rice yield changes in China

From 1980 to 1997, rice production showed a fluctuating upward trend, reaching a historical high of 200.7 million tons in 1997, an increase of 43.5% over the level in 1980. During this period, rice yield also showed a significant increasing trend, from 4,129.6 kg/hm^2 in 1980 to 6,319.4 kg/hm^2 in 1997, an increase of 53.0%. Between 1998 and 2003, rice prices experienced a dramatic drop, which directly led to a decrease in rice production and yield; rice production decreased from 198.7 million tons to 160.7 million tons, a reduction of 19.1%, while rice yield decreased from 6,366.2 kg/hm^2 to 6,060.7 kg/hm^2, representing a 4.8% drop. The rice yield decrease was relatively small, which played an important role in sustaining rice production. Without this relatively stable yield level, rice production would have witnessed a more significant decrease during this period. Since 2004, the Chinese government has implemented a price support policy for rice production, and rice prices have gradually increased. This has led to a steady increase in both rice yield and production, reaching 7,044.3 kg/hm^2 and 211.9 million tons in 2020, respectively. Generally, rice yield and production show high co-movement. The scarcity of land in China means that significant increases in rice planting areas are unlikely in the future. Since rice yield largely determines total rice production, enhancing the yield level is the only path to significantly increasing the total rice production in China (Figure 2-1).

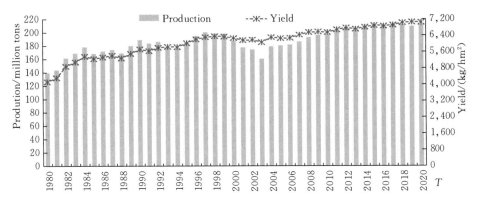

Figure 2-1　Changes in rice yield and production, 1980 – 2020

Source: National Bureau of Statistics.

2.2　Evolution in rice seed varieties in China

In terms of changes in major diffused rice varieties during the period 1980 to 2019, the diffusion period for rice seed varieties can be divided into six segments: the first period was from 1980 to 1981 when Guichao No. 2 (GC 2) was planted in vast areas; the second period ranged from 1982 to 1985 when Shanyou No. 2 (SY2) became the most diffused rice variety; the third period started from 1986 to 1994, during which Shanyou No. 63 (SY63) became the top variety in terms of diffusing areas; the fourth period ranged from 1995 to 1999 when Gangyou No. 22 (GY22) began to

CHAPTER 2   The contribution of seed variety to rice yield increases

be adopted and diffused in most rice planting areas in China; the fifth period was from 2000 to 2007, the rice leading variety was changed to Liangyoupei No. 9 (LYP9); and finally, during the sixth diffusion period from 2008 to 2019, when the leading rice seed variety was unclear as wide varieties were developed and diffused in rice production and the variety concentration rate was low (Figure 2 – 2).

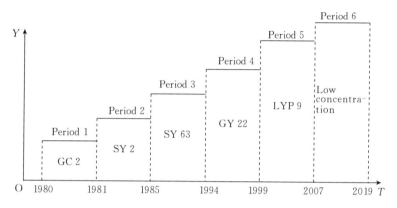

Figure 2-2   Leading varieties and diffusion period for corn, 1980 – 2019

2.3   Specification for rice yield model

Since the diffusion period for rice seed varieties can be split into six distinct periods between 1980 and 2019, according to the empirical method presented in Chapter 1, rice yield can be specified as a model by including five dummy variables representing various varieties in the basic yield response equation

$$Y_{it} = C + \beta_0 Y_{it-1} + \beta_1 P_{it-1} + \beta_2 D_2 + \beta_3 D_3 + \beta_4 D_4 + \beta_5 D_5 + \beta_6 D_6$$

(2-1)

where Y indicates rice yield level; P represents rice producer prices; i indicates the main producing provinces in China; t is the year from 1980 to 2019 and $t-1$ represents the previous year. D_2 to D_6 are dummy variables distinguishing different diffusion periods; D_2 denotes seed diffusion Period 2 (SY2), with 1982 to 1985 as 1, otherwise, 0; D_3 represents diffusion Period 3 (SY63), with 1986 to 1994 as 1, otherwise, 0; D_4 represents diffusion Period 4 (GY22), with 1995 to 1999 as 1, otherwise, 0; D_5 represents diffusion Period 5 (LYP9), with 2000 to 2007 as 1, otherwise, 0; and D_6 represents diffusion Period 6 when the variety concentration rate was low, with 2008 to 2019 as 1, otherwise, 0.

2.4 Model estimation

2.4.1 Estimation data

Provincial panel data on rice from 1980 to 2019 in China are used for estimation. Rice yields are obtained from the China Rural Statistical Yearbook; price data are drawn from the National Cost and Return of Agricultural Products in China and are expressed in real terms with 1980 as the base year; information on the major rice varieties diffused in each province is from the annual internal reports from 1980 to 2019 released by the Seed Administration of the Ministry of Agriculture and Rural Affairs of China (Table 2-1).

Table 2-1 Descriptive statistics for rice yield and prices

Variable	Mean	Std. Deviation	Minimum	Maximum
Yield	6,223.89	1,375.29	1,230.70	9,458.65
Price	18.15	4.45	11.76	26.07

CHAPTER 2   The contribution of seed variety to rice yield increases

2.4.2   Estimation results

The yield equation is a dynamic panel model. The existence of endogeneity means that the Ordinary Least Square (OLS) method does not work in estimating dynamic panel models; thus, the Arellano-Bond method, which is a type of generalized method of moments (GMM), is used to estimate the coefficients. This estimation method can correct the regression bias caused by endogeneity. The estimated results are reported in Table 2-2.

Table 2-2   Estimated results for rice yield model

Variable	Coefficients	Std. Error	Prob.
Cons.	1,955.33	298.65	0.000***
Y_{t-1}	0.52	0.05	0.000***
P_{t-1}	12.24	4.92	0.013**
D_2	498.69	104.07	0.000***
D_3	629.70	111.37	0.000***
D_4	946.12	137.00	0.000***
D_5	956.33	154.80	0.000***
D_6	1,170.72	170.28	0.000***
Obs.	669	Wald	1,253.63

Note: ***, **, and * indicate the 1%, 5%, and 10% significance levels, respectively.

The estimated results show that most coefficients on the variety dummy variables are highly significant at the 1% significance level and take positive signs, strongly suggesting that the model adequately captures the net increases in rice yield that can be associated with the replacements in various seed varieties. The parameters on the dummy variables represent the average increases in rice yields attributed to the diffusion of seed varieties over the average yield in the base period. For instance, the parameter for D_2 is 498.69, which implies that new seed diffusion in Period 2 (1982 to 1985) increased the average yield by 498.69 kg/hm^2 over the average in the base period (1980 to 1981). Likewise, in Period 3 (1986 to 1994), the average increase attributed to variety replacement in this period was 629.70 kg/hm^2, a net increase of 131.01 kg/hm^2 or 26.3% over the increase in Period 2, indicating that rice yields in this period were promoted by 629.70 kg/hm^2 relative to the base period. In Period 4 (1995 to 1999), the rise was 946.12 kg/hm^2, an increase of 316.42 kg/hm^2 or 50.2% over that in the previous period. In Period 5 (2000 to 2007), the diffusion of the new rice variety increased rice yield by 956.33 kg/hm^2, a slight increase of 10.21 kg/hm^2 or 1.1% over the increase in period 4. Finally, in Period 6 (2008 to 2019), the increase due to new rice variety diffusion jumped to a high of 1,170.72 kg/hm^2, an increase of 214.39 kg/hm^2 or 22.4% over that in Period 5. Overall, the average increase in rice yields due to the diffusion of new varieties compared to the base period rose over time, indicating a significant improvement in seed breeding and diffusion.

2.5 Empirical results

2.5.1 Absolute contribution of rice varieties

The absolute contribution is the net increase attributed to crop

CHAPTER 2　The contribution of seed variety to rice yield increases

variety replacement with a unit of kilogram per hectare. Figure 2-3 illustrates the absolute contribution of varieties to rice yield increases. The base period for rice variety diffusion is from 1980 to 1981, when the leading rice variety was Guichao No. 2; as Shanyou No. 2 became the major rice variety in Period 2, the average rice yield was increased by 498.7 kg/hm^2. In Period 3 (1986 to 1994), when Shanyou No. 63 was extensively diffused in nationwide, rice average yield was increased by 629.7 kg/hm^2, an increase of 131.01 kg/hm^2 or 26.3% over the absolute contribution made by the variety in Period 2. In the fourth period (1995 to 1999), rice yield witnessed a net increase of 946.1 kg/hm^2 over the base period yield due to the rice variety replacement, a jump of 316.4 kg/hm^2 or 50.2% over the contribution in the third period, indicating that the rice variety Gangyou No. 2 diffused in this period had significantly promoted rice yield level. In the fifth period (2000 to 2007), the net increase in yield caused by variety replacement was 956.3 kg/hm^2, only 10.2 kg/hm^2 or 1.1% higher than the contribution of variety in the previous period. The increase in the absolute contribution in Period 5 was the lowest among all variety diffusion periods, which may be related to factors such as low rice producer prices during this period. The low prices harmed farmers' enthusiasm, meaning that they were less likely to adopt new varieties and paid less attention to field management, inhibiting the yielding potential of the new varieties. Thus, crop prices may be a factor in ensuring the performance of advanced varieties on increasing crop yield levels. In Period 6 (2008 to 2019), the variety contribution to yield increase jumped to a high of 1,170.7 kg/hm^2, a net increase of 214.4 kg/hm² or 22.4% over the fifth diffusion period for rice varieties, exceeding the 1,000 kg/hm² for the first time, indicating significant progress in rice breeding technologies and seed diffusion work in this period (See Figure 2-3).

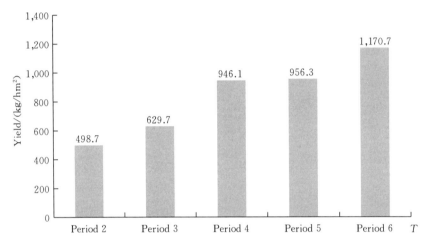

Figure 2-3 Absolute contribution of varieties to rice yield in the base period

Note: Rice varieties contain six periods: 1980 – 1981, 1982 – 1985, 1986 – 1994, 1995 – 1999, 2000 – 2007, and 2008 – 2019.

2.5.2 Contribution rate of rice varieties

The contribution rate is a relative indicator, which equals the share of the net increase attributed to variety changes to the total yield increment; a high contribution rate suggests that varieties play an important role in enhancing crop yields. Table 2-3 reports the detailed changes in rice yield, as well as the contribution of seed varieties. The yield average in the base period (1980 – 1981) was 4,538.6 kg/hm², which increased to 5,218.0 kg/hm² in Period 2

CHAPTER 2   The contribution of seed variety to rice yield increases

(1982 – 1985), an increment of 679.4 kg/hm^2; of this, 498.7 kg/hm^2 was attributed to the replacement of varieties, so the contribution rate of seed varieties in this diffusion stage over the base period equaled 73.4%. In Period 3, the average yield of rice grew to 5,633.8 kg/hm^2, an increase of 1,095.2 kg/hm^2 over the average in the base period, of which 629.7 kg/hm^2 was attributed to new seed varieties; thus, the contribution rate of seed varieties for the third diffusion period was 57.5%, nearly 16 percentage points lower than that in the second period. In Period 4, rice average yield increased to 6,465.1 kg/hm^2, 1,926.5 kg/hm^2 higher than that in the base period, while the diffusion of the new seed variety was estimated as being responsible for 956.3 kg/hm^2 of the total yield increase, resulting in a contribution rate of 49.1% for this period, 8.4 percentage points lower than the third period. In Period 5, rice average yield was 6,433.3 kg/hm^2, 31.7 kg/hm^2 lower than the fourth period, mainly owing to reduced producer prices in this period, while the variety contribution was estimated at a level of 956.3 kg/hm^2; yield average in Period 5 was 1,894.7 kg/hm^2 greater than the base period, so the contribution rate over the base period was 50.5%, 1.4 percentage points higher than the rate in the fourth period. Period 6 saw a significant increase in the absolute contribution of new rice varieties, with this jumping to 1,170.7 kg/hm^2, a rise of 214.4 kg/hm^2 or 22.4% over the previous level; the total rice yield average was increased to 6,972.9 kg/hm^2, an increase of 2,434.3 kg/hm^2 over the base period, so the contribution rate was calculated at 48.1%, 2.4 percentage points lower than that in Period 5 (Table 2-3).

Table 2-3 Varietal contribution to rice yield increases over the base period (kg/hm^2, %)

Period	Yield	Yield increase	Varietal contribution	Contribution rate
Base (1980 – 1981)	4,538.6	—	—	—
2 (1982 – 1985)	5,218.0	679.4	498.7	73.4
3 (1986 – 1994)	5,633.8	1,095.2	629.7	57.5
4 (1995 – 1999)	6,465.1	1,926.5	946.1	49.1
5 (2000 – 2007)	6,433.3	1,894.7	956.3	50.5
6 (2008 – 2019)	6,972.9	2,434.3	1,170.7	48.1

Note: yield data are obtained from the National Bureau of Statistics.

Table 2-4 reports the contribution that varieties make to rice yield increases over the previous period. The seed contribution over the previous period in Period 2 is, therefore, the same as that over the base period (73.4%) because the comparison is made between the same two periods. In Period 3, the average rice yield was 5,633.8 kg/hm^2, 415.8 kg/hm^2 higher than the yield level in Period 2, and the increase caused by seed variety replacement over Period 2 was calculated as 131.0 kg/hm^2 (from 629.7 to 498.7); thus, the contribution of new seed varieties in Period 3 over the previous period is 31.5%. In Period 4, rice average yield increased from 5,633.8 kg/hm^2 in the previous period to 6,465.1 kg/hm^2, a net increase of 831.3 kg/hm^2; notably, the net increase due to new

CHAPTER 2 The contribution of seed variety to rice yield increases

variety diffusion shrank to 316.4 kg/hm^2, although the seed contribution rate over the previous period increased to 38.1%, 6.6 percentage points higher than the rate in the previous period. In Period 5, rice yield witnessed a slight drop to 6,433.3 kg/hm^2, a net decrease of 31.7 kg/hm^2 over the previous period. Low prices during Period 5 may explain this decrease, but since the yield level was reduced, it is impossible to calculate the contribution rate of varieties, although there is still an absolute positive contribution of 10.2 kg/hm^2 for rice varieties. Finally, in Period 6, the yield increase caused by new variety diffusion jumped from the previous 956.3 kg/hm^2 to 1,170.7 kg/hm^2, a rise of 214.4 kg/hm^2. Meanwhile, the rice yield average in Period 6 was 6,972.9 kg/hm^2, an increase of 539.6 kg/hm^2 over the average in Period 5; as a result, the seed variety contribution reached 39.7%, 1.6 percentage points higher than that in the fourth diffusion period. (Table 2-4).

Table 2-4 **Varietal contribution to rice yield increases over the previous period** (kg/hm^2, %)

Period	Yield	Yield increase	Varietal contribution	Contribution rate
Base (1980 – 1981)	4,538.6	—	—	—
2 (1982 – 1985)	5,218.0	679.4	498.7	73.4
3 (1986 – 1994)	5,633.8	415.8	131.0	31.5

(续)

Period	Yield	Yield increase	Varietal contribution	Contribution rate
4 (1995 – 1999)	6,465.1	831.3	316.4	38.1
5 (2000 – 2007)	6,433.3	−31.8	10.2	—
6 (2008 – 2019)	6,972.9	539.6	214.4	39.7

Note: yield data are obtained from the National Bureau of Statistics.

2.6 Summary

The rice variety diffusion period can be divided into six stages in terms of the replacement of leading varieties. Using provincial data, this chapter constructed a rice yield model considering variety replacement to estimate the contribution of varieties to rice yield increase. According to the model estimation results, the contribution of rice varieties to rice yield increase in the current period (Period 6) was estimated at a level of 1,170.7 kg/hm^2, and the total increase in rice yield was 2,434.3 kg/hm^2 compared with the base period; therefore, the contribution rate of new rice varieties to yield increase was 48.1% (1,170.7/2,434.3). In the four historical periods, the net contribution to yield increase for varieties was 498.7 kg/hm^2, 629.7 kg/hm^2, 946.1 kg/hm^2, and 956.3 kg/hm^2, respectively, and the corresponding contribution rates of varieties to yield increase was 73.4%, 57.5%, 49.1%, and

50.5%.

The absolute contribution that rice varieties make to yield increase has been increasing steadily. A breakthrough was made in the mid-1990s, during which time, the absolute increase attributed to varieties increased from 629.7 kg/hm^2 to 946.1 kg/hm^2, a significant increase of 50.2%. At the current stage (Period 6), the net increase level of rice yield attributed to varieties has exceeded 1,000 kg/hm^2, reaching a historical high of 1,170.7 kg/hm^2, indicating that rice varieties play a pivotal role in determining rice yield. A decreasing trend can be observed regarding the contribution rate for rice varieties over the base period. The historical contribution rate of varieties reached 73.4% during the mid-1980s and decreased to 57.5% by the mid-1990s before dropping further to 48.1% in the current diffusion stage. Although the current contribution rate is close to 50%, which is still relatively high, this is notably lower than the highest recorded level of more than 70%. As mentioned above, the absolute contribution of rice varieties to yield increase continues to rise, but the reasons for the continuous decline of the contribution rate should be explained. This can be caused by the extensive use of material elements such as fertilizers and pesticides, which have significantly raised rice yield levels but reduced the contribution share made by varieties, even if the level of absolute contribution of rice varieties keeps increasing.

CHAPTER 3 The contribution of seed variety to wheat yield increases

3.1 Wheat yield changes in China

Between 1980 and 1984, the tremendous dividend of the household responsibility system was fully released, causing wheat yield and production to increase rapidly within a short period. Wheat yield increased from 1,913.89 kg/hm² in 1980 to 2,969.08 kg/hm² in 1984, with the mean annual growth rate reaching 9.18%; wheat production swiftly increased from 55.205 million tons in 1980 to 87.82 million tons in 1984, with a mean annual growth rate of 9.73%. From 1984 to 1997, under the stimulus provided by the grain protective price policy, wheat yield and production continued to maintain a significant growth trend, reaching a record high level of 4,101.87 kg/hm² and 123.29 million tons in 1997, respectively, 38.2% and 40.4% higher than that in 1984. The annual growth rates were 2.33% and 2.45%, respectively, slower than the level during the previous period.

Between 1997 and 2003, following the grain protective price policy's abolition, China witnessed a sharp decline in wheat production, while wheat yield also declined slightly. Wheat yield declined by 4.1%, from 4,101.87 kg/hm² in 1997 to 3,931.84 kg/

hm² in 2003. After the elimination of grain protective prices, farmers' enthusiasm to grow grain was frustrated, resulting in a sharp decline in the wheat planting area and a marginal decline in wheat yield. Meanwhile, production diminished from 123.2885 million tons in 1997 to 86.4885 million tons in 2003, declining 29.85% and seeing an average annual decline rate of 4.9%. From 2003 to 2020, under the support of certain policies, including the grain subsidies policy and the price support policy, wheat yield continued increasing in line with its production, at a mean annual rate of 2.1% and 2.6%, respectively. In 2020, China's wheat yield was 5,742.3 kg/hm², threefold greater than the 1980 yield level; wheat production reached 134.2538 million tons, which was 2.43 times the 1980 production level (Figure 3-1).

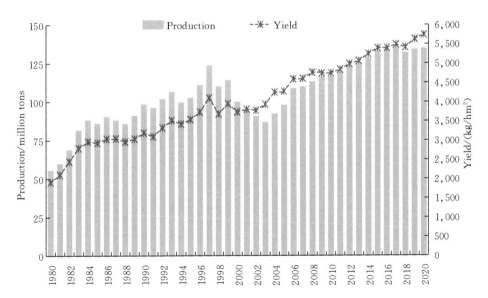

Figure 3-1 Changes in wheat yield and production, 1980-2020

Source: National Bureau of Statistics.

3.2　Evolution of wheat seed varieties in China

Regarding the replacements of major diffused varieties of wheat during the period 1980 – 2019, the wheat variety diffusion period may be distinguished into five stages. The first period spanned from 1980 to 1989; during this period, Bainong No. 3217 (BN3217) was extensively planted nationally, being the lead variety for 10 years. During the second period between 1990 and 1994, Yangmai No. 5 (YM5) replaced BN3217 as the most diffused variety. The third period ranged from 1995 to 2002, during which YM5 dropped from a lead place, and Yangmai No. 18 (YM18) became China's most diffused variety. The fourth period was from 2003 to 2008, when the major wheat variety was Zhengmai No. 9023 (ZM9023). Finally, the fifth period for wheat variety diffusion spanned between 2009 and 2019, when the top wheat variety shifted to Jimai No. 22 (JM22) (Figure 3 – 2).

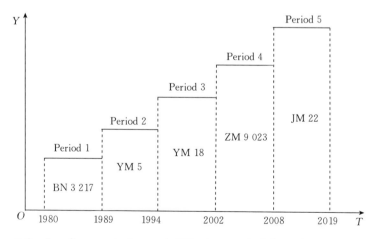

Figure 3-2　Leading varieties and diffusion period for wheat, 1980 – 2019

3.3 Specification for the wheat yield model

The variety diffusion period for wheat can be distinguished into five sub-periods between 1980 and 2019. Accordingly, the wheat yield model may be specified as an equation including four dummy variables in the basic yield response equation

$$Y_{it} = C + \beta_0 Y_{it-1} + \beta_1 P_{it-1} + \beta_2 D_2 + \beta_3 D_3 + \beta_4 D_4 + \beta_5 D_5$$

(3-1)

where Y indicates wheat yield; P represents the producer prices for wheat; i indicates China's leading wheat producing provinces; t is the year from 1980 to 2019, and $t-1$ represents the previous year, while $D_2 - D_5$ are dummy variables distinguishing different diffusion periods. D_2 denotes seed diffusion Period 2, with 1990 to 1994 as 1, otherwise, 0; D_3 represents diffusion Period 3, with 1995 to 2002 as 1, otherwise, 0; D_4 represents diffusion Period 4, with 2003 to 2008 as 1, otherwise, 0, while D_5 represents diffusion Period 5, with 2009 to 2019 as 1, otherwise, 0.

3.4 Model estimation

3.4.1 Estimation data

Provincial panel data on wheat from 1980 to 2019 in China are adopted for the estimation. Wheat yield data are obtained from the China Rural Statistical Yearbook. Wheat price data are drawn from the National Cost and Return of Agricultural Products in China, being expressed in real terms with 1980 as the base year.

Information on the major wheat varieties diffused in each province is taken from the annual internal reports between 1980 and 2019, provided by the Seed Administration of the Ministry of Agriculture and Rural Affairs of China (Table 3-1).

Table 3-1 Descriptive statistics for wheat yield and prices

Variable	Mean	Std. Deviation	Minimum	Maximum
Yield	3,236.30	1,321.51	600.00	6,898.77
Price	17.45	2.40	12.94	21.76

3.4.2 Estimation results

The Arellano-Bond method was employed to estimate the model, the estimated results are reported in Table 3-2. All variables are highly significant at the 1% significance level and take positive signs, which strongly indicates that the model sufficiently captures the wheat yield's net increases caused by the diffusion of various wheat varieties. The dummy variables' parameters represent the mean increases in wheat yields attributed to variety replacement over the average yield level during the base period.

Table 3-2 Estimated results for the wheat yield model

Variable	Coefficients	Std. Error	Prob.
C	610.23	90.84	0.000***
Y_{t-1}	0.62	0.043	0.000***
P_{t-1}	19.14	4.07	0.000***
D_2	204.00	41.84	0.000***
D_3	264.08	43.17	0.000***

CHAPTER 3　The contribution of seed variety to wheat yield increases

（续）

Variable	Coefficients	Std. Error	Prob.
D_4	464.00	56.87	0.000***
D_5	535.01	67.75	0.000***
Obs.	673	Wald	962.33

Note: ***, **, and * indicate the 1%, 5%, and 10% significance level, respectively.

The parameter for D_2 is 204.00, implying that the varieties diffused during Period 2 (1990 - 1994) increased the average wheat yield by 204.00 kg/hm² over the average during the base period (1980 - 1989). According to the estimated coefficients, during Period 3 (1995 - 2002), the mean increase was 264.08 kg/hm², a net increase of 60.08 kg/hm² or 29.5% over the increase during period 2. In Period 4 (2003 - 2008), the increase stemming from variety replacement was 464.00 kg/hm², a jump of 199.92 kg/hm² or 75.7% over that in the previous period. In Period 5 (2009 - 2019), the increase jumped further to a new high of 535.01 kg/hm², which was 71.01 kg/hm² or 15.3% higher than the increase during Period 4.

3.5　Empirical results

3.5.1　Absolute contribution of wheat varieties

Figure 3-3 illustrates the varieties' absolute contribution to wheat yield increases. The base period for wheat varieties diffusion is from 1980 to 1989 when the leading wheat variety was Bainong No. 3217. Yangmai No. 5 became the major wheat variety during the

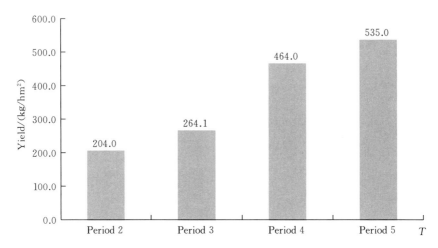

Figure 3-3 Absolute contribution of variety to wheat yield in the base period

Note: wheat varieties contain five periods, which are 1980 - 1989, 1990 - 1994, 1995 - 2002, 2003 - 2008, and 2009 - 2019, respectively.

second period (1990 - 1994) when the mean wheat yield increased by 204.0 kg/hm². In period three (1995 - 2002), when Yangmai No. 18 was extensively diffused nationwide, the average wheat yield increased by 264.1 kg/hm², which was 60.1 kg or 29.5% over the absolute contribution made by the variety during period two. In the fourth period (2003 - 2008), wheat yield witnessed a net increase of 464.0 kg/hm² over the base period, due to the dominant wheat variety being replaced by Zhengmai No. 9023. This achieved a jump of 199.9 kg/hm² or 75.7% over the contribution in the third period, marking the highest increase in the absolute contribution among all of the variety diffusion periods. This establishes that the diffused wheat variety Zhengmai No. 9023 significantly promoted the wheat yield level in this period. In the fifth period (2009 - 2019), the leading wheat variety shifted to Jimai No. 22,

CHAPTER 3 The contribution of seed variety to wheat yield increases

with the wheat yield seeing a net increase of 535.0 kg/hm² due to variety replacement, which was 71.0 kg/hm² or 15.3% higher than the variety's contribution during the prior period. Overall, the mean increase in wheat yields due to the diffusion of new varieties over the base period increased over time, indicating significant improvements in terms of seed breeding and diffusion.

3.5.2 Contribution rate of wheat varieties

Table 3-3 reports the detailed changes in wheat yield and the contribution rates of varieties to yield increases over the base period. The mean yield during the base period (1980 – 1989) was 2,390.8 kg/hm², which increased to 2,965.6 kg/hm² in Period 2 (1990 – 1994), an improvement of 574.8 kg/hm²; of this, new varieties contributed 204.0 kg/hm² to the total increase, meaning the new seed varieties' contribution equaled 35.5% during this diffusion period. In Period 3, the mean wheat yield expanded to 3,231.5 kg/hm², an increase of 840.7 kg/hm² over the average in the base period, of which 264.1 kg/hm² was attributed to new seed varieties' diffusion. Therefore, the new seeds' contribution during the third diffusion period was 31.4%. This period witnessed a slight decline in the proportion of improvement attributed to new seed diffusion, with the estimated contribution rate 4.1 percentage points lower compared with the previous period. In Period 4, the average wheat yield was 3,557.8 kg/hm², 1,167.0 kg/hm² higher than in the base period, while new seed diffusion was estimated as accounting for 464.0 kg/hm² of the total increase. This resulted in a contribution rate of 39.8% for this period, 8.4 percentage points higher than the previous period. Subsequently, a significant increase in new seed varieties' contribution was seen during the fifth diffusion period, with this increase being due to new seed diffusion jumping from 1,167.0 kg/hm² to 1,583.7 kg/hm², a net increase of 416.7 kg/hm² or 35.7%. The total increase in wheat yield average was 3,974.6 kg/hm², marking a jump of 1,583.7 kg/hm² from the base period, although the contribution rate increased to 33.78%, which was 5.98 percentage points lower compared with the fourth period. Generally, the varieties' contribution to expanded wheat yields did not see obvious variations, which bonded between 30% and 40% and was significantly lower than the contribution made by rice varieties.

Table 3-3　Varietal contribution to wheat yield increases over the base period（kg/hm², %）

Period	Yield	Yield increase	Varietal contribution	Contribution rate
Base (1980–1989)	2,390.8	—	—	—
2 (1990–1994)	2,965.6	574.8	204.0	35.5
3 (1995–2002)	3,231.5	840.7	264.1	31.4
4 (2003–2008)	3,557.8	1,167.0	464.0	39.8
5 (2009–2019)	3,974.6	1,583.8	535.0	33.8

Note: Yield data are obtained from the National Bureau of Statistics.

CHAPTER 3 The contribution of seed variety to wheat yield increases

Table 3-4 reports the detailed changes in wheat yield and the contribution rates of varieties to yield increases over the previous period. Because the comparison is made between the same two periods, the variety contribution in Period 2 over the previous period was equivalent to that over the base period (35.5%). In period three, the mean wheat yield was 3,231.5 kg/hm^2, 265.9 kg/hm^2 higher than the previous period, while the increase as a consequence of variety replacement over Period 2 was estimated as being 60.1 kg/hm^2; accordingly, the contribution of seed varieties during Period 3 compared with the previous period was 22.60%. In the fourth period, the average wheat yield exhibited a significant increase, from 3,232.5 kg/hm^2 in the previous period to 3,557.8 kg/hm^2, a net increase of 326.3 kg/hm^2, while the increase stemming from new variety diffusion expanded to 199.9 kg/hm^2. This caused the variety contribution rate to 61.3%, 38.7 percentage points greater than the previous period. Finally, a significant decline in absolute variety contribution was witnessed in the fifth diffusion period. The increase attributed to new variety diffusion dropped from 199.9 kg/hm^2 to 71.0 kg/hm^2, a reduction of 128.9 kg/hm^2 or 64.5%. Meanwhile, the average wheat yield in this period was 3,974.6 kg/hm^2, an increase of 416.8 kg/hm^2 or 11.7% over the mean in period four. Consequently, in this period, variety contribution reached a low level of 17.0%, 44.23 percentage points below that during the fourth diffusion period (Table 3-4).

Table 3-4 Varietal contribution to wheat yield increases over the previous period (kg/hm^2, %)

Period	Yield	Yield increase	Varietal contribution	Contribution rate
Base (1980 – 1989)	2,390.8	—	—	—
2 (1990 – 1994)	2,965.6	574.8	204.0	35.5
3 (1995 – 2002)	3,231.5	265.9	60.1	22.6
4 (2003 – 2008)	3,557.8	326.3	199.9	61.3
5 (2009 – 2019)	3,974.6	416.8	71.0	17.0

Note: yield data are obtained from the National Bureau of Statistics.

3.6 Summary

Wheat varieties diffused between 1980 and 2019 in China can be divided into five sub-periods. During the current diffusion period, wheat varieties contributed 535.0 kg/hm^2 over the varieties diffused in the based period, while the contribution rate was 33.8%. The wheat yield benefit from wheat varieties during each historical period increased to 204.0 kg/hm^2, 264.1 kg/hm^2, and 464.0 kg/

CHAPTER 3 The contribution of seed variety to wheat yield increases

hm^2 for Period 2 (1990 – 1994), Period 3 (1995 – 2002), and Period 4 (2003 – 2008), respectively; with respective contribution rates being 35.5%, 31.4% and 39.8%. Although the wheat yield benefit of varieties increased from 204.0 kg/hm^2 during the early 1990s to 535.0 kg/hm^2 in the current stage, wheat varieties' contribution rate was merely 33.8%, which was significantly lower than the contribution of rice varieties. Even the record high level was under 40%, evidencing that significant improvement was not in place for breeding good wheat varieties in China, with no breakthrough of nationwide varieties achieved in a long time.

The potential reasons for the low contribution of wheat varieties to yield increases could be summarized as follows.

First, the upgrading of major wheat varieties was relatively slow. Most wheat varieties are conventional varieties (not hybrid varieties), and wheat products of these varieties can be persistently used as seeds for the next production; this may lead to slow updating of wheat varieties in actual wheat production, leading to clear degradation of varietal characteristics and further restricting increases in wheat yields.

Second, there was an insufficient investment in the research and development of wheat varieties. For a long time, affected by the challenges of cultivating wheat varieties, particularly hybrid varieties, the Chinese seed industry's research and development focused on rice and corn. Insufficient financial investment in wheat variety breeding restricted the vitality of scientific and technological innovation in wheat breeding, leading to a shortage of dominant varieties making a significant national impact.

Third, the enthusiasm for using wheat varieties was low. Out of the three main grain crops, wheat's planting income was relatively low, resulting in limited enthusiasm among farmers to plant wheat and a low willingness to adopt new varieties of wheat, thereby undermining wheat varieties' contribution to yield increase.

CHAPTER 4 The contribution of seed variety to corn yield increases

4.1 Corn yield changes in China

China's corn production has maintained its rapid and stable growth since the 1980s. Production expanded from 62.60 million tons in 1980 to 260.67 million tons in 2020, representing an average annual growth rate of 3.63% or an expansion of over 400% in the last 40 years. China's corn yield has also shown an upward trend, from 3,116.4 kg/hm^2 in 1980 to 6,317.0 kg/hm^2 in 2020. This represents an average annual growth rate of 1.78%, meaning yields have almost doubled in 40 years.

The change in corn yield and production can be divided into three phases. The first phase was from 1980 to 2003, when corn yield and production were volatile. During this period, corn yield grew by an average of 2.71% per year, while production grew by 1.91%. During this period, yield and production decreased significantly in 1997 and 2000. Production decreased by 23.16 million tons and 22.09 million tons respectively, while yield decreased by 816.0 kg/hm^2 and 347.2 kg/hm^2 compared with the previous year, representing the biggest fall since 1980. The second phase was from 2004 to 2015, when corn yield entered a period of stable and rapid growth. During this time, agriculture underwent considerable modernization. The promotion and use of improved varieties and agricultural machinery significantly improved corn yields. Corn production increased from 115.83 million tons to 264.99 million tons in 12 years, with a growth rate of 6.64% per year. This is much higher than the increased rate of production in the first phase. However, corn yield increased from 5,120.2 kg/hm^2 to 5,892.9 kg/hm^2 during this phase, with an average annual growth of 1.29%, which was lower than the previous phase. The third phase was from 2016 to 2020, when corn production decreased slightly to 266.7 million tons in 2020. Meanwhile, corn yield increased to 6,317.0 kg/hm^2 in 2020 with an average annual growth of 1.43%, higher than the previous phase. This shows that China's agricultural supply-side structural reform significantly effected these years. Corn's production structure and production mode were constantly improved, so that corn production was more in line with market demand and avoided excess production capacity (Figure 4-1).

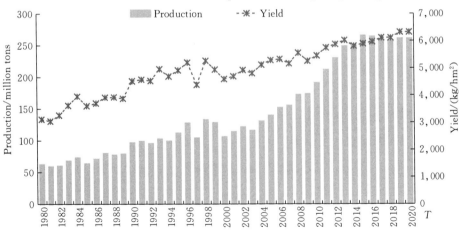

Figure 4-1 Changes in corn yield and production, 1980–2020
Source: National Bureau of Statistics.

CHAPTER 4　The contribution of seed variety to corn yield increases

4.2　Evolution in corn seed varieties in China

According to the changes in major diffused corn varieties during the period 1980 to 2019, the diffusion period for corn seed varieties can be divided into five segments: the first period was from 1980 to 1986, when Zhongdan No. 2 (ZD 2) was planted in extremely large areas; the second period was from 1987 to 1994 when Danyu No. 13 (DY 13) replaced ZD 2 as the most diffused corn variety; the third period ranged from 1995 to 1999, during which DY 13 was dropped from first place and Yedan No. 13 (YD 13) began to be adopted and diffused in most corn regions in China; the fourth period was from 2000 to 2003, the major variety was changed to Nongda No. 108 (ND 108) in this diffusion period; and finally, the fifth period was from 2004 to 2019 when the top corn seed variety was Zhengdan No. 958 (ZD 958), which became the most popular corn variety and took the leading place for corn variety of the nation for more than 15 years (Figure 4-2).

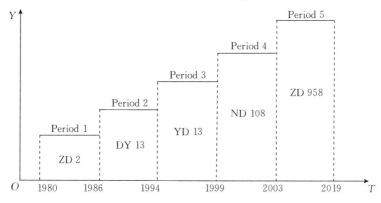

Figure 4-2　Leading varieties and diffusion period for corn, 1980 – 2019

4.3 Specification for the corn yield model

Since corn variety diffusion can be split into five stages in terms of replacements in the major diffused varieties during the period 1980 to 2019, thus corn yield can be specified as a model by including four dummy variables into the basic yield response equation

$$Y_{it} = C + \beta_0 Y_{it-1} + \beta_1 P_{it-1} + \beta_2 D_2 + \beta_3 D_3 + \beta_4 D_4 + \beta_5 D_5$$

(4-1)

where Y indicates corn yield; P represents corn producer prices; i indicates the main producing provinces in China; t is the year from 1980 to 2019; $t-1$ represents the previous year, lagged; and D_2 - D_5 are dummy variables distinguishing different diffusion periods. D_2 denotes seed diffusion Period 2, with 1987 to 1994 as 1, otherwise, 0; D_3 represents diffusion Period 3, with 1995 to 1999 as 1, otherwise, 0; D_4 represents diffusion Period 4, with 2000 to 2003 as 1, otherwise, 0; and D_5 represents diffusion Period 5, with 2004 to 2019 as 1, otherwise, 0.

4.4 Model estimation

4.4.1 Estimation data

The estimation uses provincial panel data for corn from 1980 to 2019 in China. Corn yield is obtained from the China Rural Statistical Yearbook; price data are drawn from the National Cost and Return of Agricultural Products in China and expressed in real

CHAPTER 4 The contribution of seed variety to corn yield increases

terms with 1980 as the base year; the annual reports of the Seed Administration of the Ministry of Agriculture of China from 1980 to 2019 are the source of information on the major corn varieties diffused in each province (Table 4-1).

Table 4-1 Descriptive statistics for corn yield and prices

Variable	Mean	Std. Deviation	Minimum	Maximum
Yield	4,586.22	1,454.43	1,034.50	8,825.91
Price	14.21	2.94	10.30	20.98

4.4.2 Estimation results

The Arellano-Bond estimation method is used to estimate the coefficients to correct the regression bias caused by endogeneity, and the estimated results are reported in Table 4-2. The estimated results show that all estimates on dummy variables are highly significant at the 1% significance level and take positive signs, strongly suggesting that the model adequately captures the net increases in corn yield due to the diffusion of different seed varieties. The parameters on the dummy variables represent the average increases in corn yields attributed to new seed diffusion over the average yield in the base period.

Table 4-2 Estimated results for corn yield model

Variable	Coefficients	Std. Error	Prob.
Cons.	1,557.15	266.76	0.000***
Y_{t-1}	0.49	0.069	0.000***

(续)

Variable	Coefficients	Std. Error	Prob.
P_{t-1}	6.52	6.10	0.285
D_2	393.17	63.67	0.000***
D_3	678.78	101.77	0.000***
D_4	725.17	99.93	0.000***
D_5	1,077.41	131.61	0.000***
Obs.	678	Wald	854.07

Note: ***, **, and * indicate the 1%, 5%, and 10% significance level, respectively.

The parameter for D_2 is 393.17, which implies that new seed diffusion in period two (1987 – 1994) increased average yields by 393.17 kg/hm² over the average in the base period (1980 – 1986). Likewise, the average increase in Period 3 (1995 – 1999) was 678.78 kg/hm², a net increase of 285.61 kg/hm² or 72.6% over the increase in Period 2. This indicates that corn yields in this period increased by 678.78 kg/hm² relative to the base period. Yields increased by 725.17 kg/hm² in Period 4 (2000 – 2003), an increase of 46.39 kg/hm² or 6.8% over the previous period. Finally, new seed diffusion saw yields jump to a high of 1,077.41 kg/hm² in Period 5 (2004 – 2019), an increase of 352.24 kg/hm² or 48.6% over Period 4. Overall, the average increase in corn yields due to the diffusion of new varieties rose dramatically over time, indicating a significant improvement in corn seed breeding and diffusion.

4.5 Empirical results

4.5.1 Absolute contribution of corn varieties

Figure 4-3 illustrates the absolute contribution of varieties to corn yield increases. The base period for corn variety diffusion is from 1980 to 1986 when the leading corn variety was Zhongdan No. 2 In Period 2 (1987 – 1994), as Danyu No. 13 became the main corn variety, average corn yields increased by 393.2 kg/hm^2. In Period 3 (1995 – 1999), when Yedan No. 13 was extensively diffused nationwide, average corn yields increased by 678.8 kg/hm^2, an increase of 285.6kg or 72.6% over the absolute contribution made by the variety in Period 2. This indicates that the corn variety Yedan No. 13 diffused in this period significantly promoted the corn yield. In the fourth period (2000 – 2003), corn yields witnessed a net increase of 725.2 kg/hm^2 over yields in the base period due to the introduction of Nongda No. 108, a jump of only 46.4 kg/hm^2 or 6.8% over the contribution in the third period. This increase in the absolute contribution was the lowest among all diffusion periods. In the fifth period (2004 – 2019), the net increase in yield caused by the new variety of Zhengdan No. 958 reached a high of 1,077.4 kg/hm^2, 352.2 kg/hm^2 or 48.6% higher than the variety's contribution in the previous period, indicating significant progress in corn breeding technologies and seed diffusion work occurred in this period.

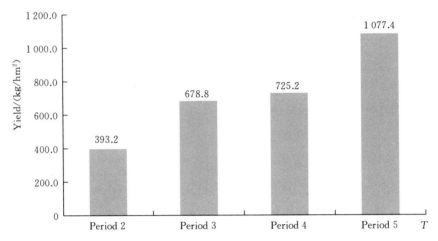

Figure 4-3   Absolute contribution of variety to corn yield over the base period

Note: corn varieties contain five periods, which are 1980 – 1986, 1987 – 1994, 1995 – 1999, 2000 – 2003, 2004 – 2019, respectively.

4.5.2 Contribution rate of seed varieties

Table 4-3 details the changes in corn yield and the contribution of new varieties. The average yield in the base period (1980 – 1986) was 3,242.0 kg/hm^2, which increased to 3,906.9 kg/hm^2 in Period 2 (1987 – 1994), an improvement of 664.9 kg/hm^2. New varieties contributed 393.2 kg/hm^2 of this to the total increase, so the new seed varieties contributed 59.1% of the increase in this diffusion period. In Period 3, the average corn yield grew to 4,564.0 kg/hm^2, an increase of 1,322.0 kg/hm^2 over the average in the base period, of which 678.8 kg/hm^2 was attributed to new seed varieties. The contribution of new seeds for the third diffusion period was thus 51.4%. In Period 4, the average corn yield was 4,706.6 kg/hm^2, 1,464.6 kg/hm^2 higher than the base period, and

CHAPTER 4   The contribution of seed variety to corn yield increases

Table 4-3   **Varietal contribution to corn yield increases over the base period** (kg/hm², %)

Period	Yield	Yield increase	Varietal contribution	Contribution rate
Base (1980–1986)	3,242.0	—	—	—
2 (1987–1994)	3,906.9	664.9	393.2	59.1
3 (1995–1999)	4,564.0	1,322.0	678.8	51.4
4 (2000–2003)	4,706.6	1,464.6	725.2	49.5
5 (2004–2019)	5,462.3	2,220.3	1,077.4	48.6

Note: yield data are obtained from the National Bureau of Statistics.

new seed diffusion was estimated as being responsible for 725.2 kg/hm² of the total increase, resulting in a contribution rate of 49.5% for this period. A slight decrease was seen in this period in the proportion of improvement attributed to new seed diffusion, which was estimated 1.9 percentage points lower than in the previous period. The contribution of new seed varieties continued to decline slightly in the fifth diffusion period, with an increase due to new seed diffusion jumping from 725.2 kg/hm² to 1,077.4 kg/hm², a net increase of 352.2 kg/hm² or 48.6%. The average corn yields increased to 5,462.3 kg/hm², a jump of 2,220.3 kg/hm² from the base period. Meanwhile, the contribution rate fell to 48.5%, 1 percentage point lower than that in the fourth period. The phenomenon of four consecutive declines was caused by a larger increase attributed to input factors, which overshadowed the contribution made by seed varieties and led to a decrease in relative variety contribution in the whole period. These estimated results suggest that new seed varieties significantly contributed to rises in corn yields, and that seed technologies and seed diffusion developments succeeded in causing significant progress in China over the past four decades.

The variety's contribution over the previous period for Period 2 is 59.1%, the same as over the base period. This is because the same two periods are being compared. The average corn yield was 4,564.0 kg/hm^2 in Period 3,657.1 kg/hm^2 higher than the previous period, and the increase caused by seed replacement over Period 2 was estimated as 285.6 kg/hm^2. The contribution of new seed varieties in Period 3 over the previous period was thus 43.5%. In the fourth period, average corn yields only increased slightly, from 4,564.0 kg/hm^2 in the previous period to 4,706.6 kg/hm^2, a net increase of only 142.6 kg/hm^2. The increase due to new seed diffusion shrank to 46.4 kg/hm^2, causing the seed contribution rate to drop to 32.5%, 11 percentage points less than in the previous period. Finally, a significant increase occurred in seed contribution in the fifth diffusion period. The increase due to new seed diffusion jumped from the previous 142.6 kg/hm^2 to 721.7 kg/hm^2, a rise of 579.1 kg/hm^2 or 406.1%. Meanwhile, the corn yield average in this period was 5,462.3 kg/hm^2, an increase of 755.7 kg/hm^2 or 16.1% over the average in Period 4. As a result, seed contribution in this period reached a high of 48.8%, 16.3 percentage points more than during the fourth diffusion period, indicating that newly

CHAPTER 4 The contribution of seed variety to corn yield increases

diffused varieties significantly boosted corn yields (Table 4-4).

Table 4-4 **Varietal contribution to corn yield increases over the previous period** (kg/hm^2, %)

Period	Yield	Yield increase	Varietal contribution	Contribution rate
Base (1980–1986)	3,242.0	—	—	—
2 (1987–1994)	3,906.9	664.9	393.2	59.1
3 (1995–1999)	4,564.0	657.1	285.6	43.5
4 (2000–2003)	4,706.6	142.6	46.4	32.5
5 (2004–2019)	5,462.3	721.7	352.2	48.8

Note: yield data are obtained from the National Bureau of Statistics.

4.6 Summary

There were five periods for corn variety diffusion between 1980 and 2019. In the current diffusion period, the varieties contributed 1,077.4 kg/hm^2 to corn yield increases, and the contribution rate was 48.6%. Corn varieties contributed 393.2 kg/hm^2, 678.8 kg/hm^2, and 725.2 kg/hm^2, respectively, to corn yield increases in Period 2, 3, and 4, and the respective contribution rates were 59.1%, 51.4%, and 49.5%. The absolute contribution increased gradually in each corn variety diffusion period. Since 2004, the first corn variety upgrading phase has greatly improved average corn yields in China. The contribution has increased from 725.2 kg/hm^2 in the previous phase (the fourth period) to 1,077.4 kg/hm^2 in the current phase (the fifth period). The net increase attributed to improved varieties was over 1,000 kg/hm^2 for the first time. The contribution rate of improved varieties was high from the mid 1980s to the mid 1990s, reaching nearly 60%. Since the mid-1990s, the corn yield contribution rate has increased stably by between 48% and 52%. Among the three major grain crops, the contribution rate of improved corn varieties was the highest, indicating that the breeding technology of corn varieties in China has made rapid progress.

CHAPTER 5 The contribution of seed variety to soybean yield increases

5.1 Soybean yield variation in China

From 1980 to 2020, there was an upward trend in soybean production and yield, with production increasing from 7.94 million tons to 19.60 million tons, and yield increasing from 1,098.8 kg/hm² to 1,983.5 kg/hm². However, these increases of only 11.66 million tons and 884.7 kg/hm², respectively, over 40 years, thus yielded average annual growth rates of only 2.28% and 1.49%. The production and yield of soybean were not significantly improved at that time due to the slow growth rates and low production levels. The soybean yield and production trend can be divided into four phases across those decades. The first phase was from 1980 to 1994, when soybean production and yield showed overall upward trends, with average annual growth rates of 5.13% and 3.32%, respectively. In 1994, soybean production increased to 16 million tons, and yield increased to 1,734.9 kg/hm². The second phase, from 1995 to 2004, was a stagnant period for Chinese soybean production, however, there was no significant growth in soybean production and yield in that decade, with average annual growth rates of only 2.86% and 0.99%, respectively. In the third phase, from 2005 to 2013, soybean production declined to 12.41 million tons, an average annual decline of 3.39%, though this was subject to a rebound in 2008; soybean yield was 1,704.5 kg/hm² in 2005 and 1,759.9 kg/hm² in 2013, an average annual growth rate of just 0.4%, showing no significant change overall. In the fourth phase, from 2014 to 2020, however, soybean production and yield showed upward trends, with average annual growth rates of 7.51% and 1.75%, respectively. During that seven years, soybean production increased from 12.69 million tons to 19.6 million tons, and yield increased from 1,787.3 kg/hm² to 1,983.5 kg/hm². Compared with that of yield, the production growth was more obvious, indicating that with the support of national policies, the planted area of soybeans expanded. Thus production increased significantly, while yield required further improvement (Figure 5-1).

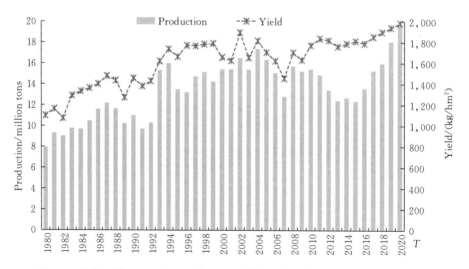

Figure 5-1 Changes in soybean yield and production, 1980－2020

Source: National Bureau of Statistics.

CHAPTER 5 The contribution of seed variety to soybean yield increases

5.2 Evolution in soybean seed varieties in China

Between 1980 and 2019, variety diffusion for soybean experienced five periods. The first period was from 1980 to 1986, during which the leading variety was Yuejin No. 5 (YJ5), and the second period lasted from 1987 to 1994, when Hefeng No. 25 (HF25) became the leading soybean variety nationwide. The third period ranged from 1995 to 1999, during which Hefeng No. 35 (HF35) began to be diffused in most soybean-producing regions in China, while the fourth period was from 2000 to 2005 when the major soybean variety changed to Suinong No. 14 (SN14). Finally, from 2006 to 2019, the fifth period has seen the previous top soybean seed varieties being replaced by Zhonghuang No. 13 (ZH13) (Figure 5-2).

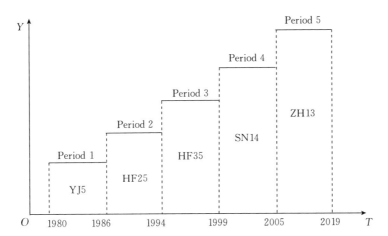

Figure 5-2 Leading varieties and diffusion period for soybean, 1980–2019

5.3 Specification for soybean yield model

Due to the diffusion period for soybean seed varieties being divided into five distinct periods between 1980 and 2019, yield can be specified as a model by including four dummy variables into the basic yield response equation:

$$Y_{it} = C + \beta_0 Y_{it-1} + \beta_1 P_{it-1} + \beta_2 D_2 + \beta_3 D_3 + \beta_4 D_4 + \beta_5 D_5$$

(5-1)

where Y indicates soybean yield; P represents soybean producer prices; i indicates the main producing provinces in China; t is the year (from 1980 to 2019), so that $t-1$ represents the previous year, lagged; and $D_2 - D_5$ are dummy variables distinguishing different diffusion periods. D_2 denotes a date being in seed diffusion period two, from 1987 to 1994, as 1, otherwise it takes the value 0; similarly, D_3 represents diffusion period three, with 1995 to 1999 as 1, otherwise, 0; D_4 represents diffusion period four, with 2000 to 2003 as 1, otherwise, 0; and D_5 represents diffusion period five, with 2004 to 2019 as 1, otherwise, 0.

5.4 Model estimation

5.4.1 Estimation data

Provincial panel data from 1980 to 2019 in China were used for estimation, with data on soybean yields obtained from the China Rural Statistical Yearbook. Price data were drawn from the National Cost and Return of Agricultural Products in China and

CHAPTER 5   The contribution of seed variety to soybean yield increases

these are expressed in real terms using 1980 as the base year. Information on major varieties diffused in each province was drawn from the annual internal reports (1980 – 2019) provided by the Seed Administration of the Ministry of Agriculture and Rural Affairs of China (Table 5-1).

Table 5-1   Descriptive statistics for soybean yield and prices

Variable	Mean	Std. Deviation	Minimum	Maximum
Yield	1,734.13	608.85	420.17	4,118.94
Price	33.68	5.27	24.24	45.92

5.4.2   Estimation results

The estimated results, based on the Arellano-Bond method, are reported in Table 5-2. The estimated results show that all estimates on dummy variables are highly significant at the 1% significance level and take positive signs, strongly suggesting that the model adequately captures the net increases in soybean yield due to the diffusion of different seed varieties, as the parameters on the dummy variables represent the average increases in soybean yields that can be attributed to new variety diffusion over the average yield in the base period.

Table 5-2   Estimated results for soybean yield model

Variable	Coefficients	Std. Error	Prob.
Cons.	515.91	106.59	0.000***
Y_{t-1}	0.57	0.06	0.000***

(续)

Variable	Coefficients	Std. Error	Prob.
P_{t-1}	1.22	1.61	0.285
D_2	71.99	26.72	0.000***
D_3	177.13	46.37	0.000***
D_4	236.01	49.32	0.000***
D_5	329.37	65.65	0.000***
Obs.	655	Wald	723.74

Note: ***, **, and * indicate the 1%, 5%, and 10% significance level, respectively.

The parameter for D_2 is 71.99, which implies that soybean variety diffused in Period 2 (1987 – 1994) increased the average yield of soybeans by 71.99 kg/hm² over the average in the base period (1980 – 1986). Likewise, in Period 3 (1995 – 1999), the average increase was 177.13 kg/hm², a net increase of 105.14 kg/hm² or 146% over the increase in Period 2, indicating that soybean yields in this period were increased by 177.13 kg/hm² relative to the base period. In Period 4 (2000 – 2005), the rise was 236.01 kg/hm², an increase of 58.88 kg/hm² or 33.3% over the yield in the previous period. Finally, in Period 5 (2006 – 2019), the increase due to new seed diffusion jumped to a high of 329.37 kg/hm², an increase of 93.36 kg/hm² or 39.6% over that in Period 4. Overall, the average increase in soybean yields due to the diffusion of new varieties over the base period rose dramatically, indicating significant improvements in seed breeding and diffusion.

CHAPTER 5 The contribution of seed variety to soybean yield increases

5.5 Empirical results

5.5.1 Absolute contribution of soybean varieties

The absolute contribution is calculated using the net increase attributed to a crop variety replacement in units of kilograms per hectare. Figure 5-3 illustrates the absolute contribution of various varieties to soybean yield increases. The base period for soybean varieties diffusion was from 1980 to 1986, when the leading soybean variety was Yuejin No. 5; in Period 2 (1987 – 1994), as Hefeng No. 25 became the leading soybean variety, the average soybean yield was increased by 72.0 kg/hm^2, while in Period 3 (1995 – 1999), when Hefeng No. 35 was extensively diffused nationwide, the average soybean yield was increased by 177.1 kg/hm^2, an increase of 105.1 kg or 146% over the absolute contribution made by the variety change in period two. In Period 3, the increase in the absolute contribution was also the highest among all the different variety diffusion periods. In the fourth period (2000 – 2005), as the leading variety changed to Suinong No. 14, soybean yields witnessed a net increase of 236.0 kg/hm^2 over the yield in the base period attributable to soybean variety replacement, which was 58.9 kg/hm^2 or 33.3% higher than the contribution in the third period. In the fifth period (2006 – 2019), the net increase in yield caused by the new variety distribution, that of Zhonghuang No. 13, was 329.4 kg/hm^2, a figure 93.4 kg/hm^2 or 39.6% higher than the contribution of variety seen in the previous period. The increase in the absolute contribution in the fifth period was the second highest among all of the variety diffusion periods, indicating significant progress in soybean breeding technologies and seed diffusion in this period.

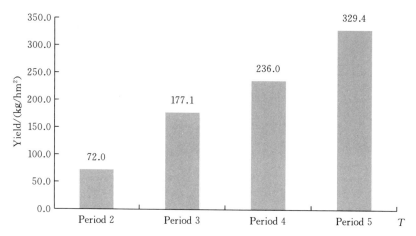

Figure 5-3 Absolute contribution of variety to soybean yield in the base period

Note: soybean varieties contain five periods, which are 1980 – 1986, 1987 – 1994, 1995 – 1999, 2000 – 2005, and 2006 – 2019, respectively.

5.5.2 Contribution rate of soybean varieties

Table 5-3 reports the detailed changes in soybean yield to clarify the contribution of the new varieties. The yield average in the base period (1980 – 1986) was 1,296.9 kg/hm^2, and this increased to 1,480.2 kg/hm^2 in Period 2 (1987 – 1994), an improvement of 183.3 kg/hm^2; of this, the new variety contributed 72.0 kg/hm^2 to the total increase, and the contribution of the new seed varieties was thus 39.3% of the increase in this diffusion period. In Period 3, the average soybean yield grew to 1,714.5 kg/hm^2, an increase of 417.6 kg/hm^2 over the average in the base period, of which 177.1

CHAPTER 5 The contribution of seed variety to soybean yield increases

kg/hm² was attributed to the new seed variety. Thus, the contribution of new seeds for the third diffusion period was 42.4%. In Period 4, the average soybean yield was 1,844.2 kg/hm², 547.3 kg higher than the base period, and new seed diffusion was estimated as being responsible for 236.0 kg/hm² of the total increase, resulting in a contribution rate of 43.1% for this period, 0.7 percentage points higher than the previous period. There was a further slight increase in the contribution of new seed varieties in the fifth diffusion period, with this increase attributable to new seed diffusion jumping from 236.0 kg/hm² to 329.4 kg/hm², a net increase of 93.4 kg/hm² or 39.6%. The total increase in soybean yield average increased to 1,992.5 kg/hm², a jump of 695.6 kg/hm² from the base period, causing the contribution rate to increase to 47.4%, 4.3 percentage points higher than that seen in the fourth period. The contribution of new seed varieties shows four distinct increases, and these estimated results show that new seed varieties significantly contribute to increases in soybean yields, suggesting that development in seed technologies and seed diffusion has made significant progress in China over the decades under consideration.

Table 5-3　Varietal contribution to soybean yield increases over the base period (kg/hm², %)

Period	Yield	Yield increase	Varietal contribution	Contribution rate
Base (1980-1986)	1,296.9	—	—	—

THE CONTRIBUTION OF SEED VARIETY TO CROP YIELD INCREASES IN CHINA

(续)

Period	Yield	Yield increase	Varietal contribution	Contribution rate
2 (1987-1994)	1,480.2	183.3	72.0	39.3
3 (1995-1999)	1,714.5	417.6	177.1	42.4
4 (2000-2005)	1,844.2	547.3	236.0	43.1
5 (2006-2019)	1,992.5	695.6	329.4	47.4

Note: yield data are obtained from the National Bureau of Statistics.

The seed contribution over the previous period for Period 2 is, of course, the same as that over the base period at 39.3%. In Period 3, the average soybean yield was 1,714.5 kg/hm^2, 234.3 kg/hm^2 higher than the previous period, and the increase caused by seed replacement over Period 2 is thus estimated as 105.1 kg/hm^2, the contribution of new seed varieties in Period 3 over the previous period was thus 44.9%. In the fourth period, the average soybean yield only exhibited a slight increase, from 1,714.5 kg/hm^2 in the previous period to 1,844.2 kg/hm^2, a net of only 129.7 kg/hm^2, causing the increase due to new seed diffusion to shrink to 58.9 kg/hm^2 and the seed contribution rate to drop to 45.4%, 0.5 percentage points more than in the previous period. Finally, in the fifth diffusion period, a significant increase in seed contribution was observed, with this increase from the previous 58.9 kg/hm^2 to 93.4 kg/hm^2, a rise of 34.5 kg/hm^2 or 58.6%. The soybean yield average in this period was 1,992.5 kg/hm², an increase of 148.3 kg/hm² or 8% over the average in Period 4. As a result, variety contribution in this period reached a high of 62.9%, 17.6 percentage points greater than that seen during the fourth diffusion period, indicating that newly diffused varieties promote soybean yield significantly, such that their contributory share exceeds the sum of all other factors driving soybean yield (Table 5-4).

Table 5-4　Varietal contribution to soybean yield increases over the previous period (kg/hm², %)

Period	Yield	Yield increase	Varietal contribution	Contribution rate
Base (1980–1986)	1,296.9	—	—	—
2 (1987–1994)	1,480.2	183.3	72.0	39.3
3 (1995–1999)	1,714.5	234.3	105.1	44.9
4 (2000–2005)	1,844.2	129.7	58.9	45.4
5 (2006–2019)	1,992.5	148.3	93.4	63.0

Note: yield data are obtained from the National Bureau of Statistics.

5.6　Summary

Soybean varieties have experienced five diffusion phases in the last 40 years. In the current stage, the contribution of seed varieties to soybean yield increases has reached 329.4 kg/hm^2, with a contribution rate of 47.4%; the contributions of seed varieties to soybean yield increases in the previous diffusion phases were 72.0 kg/hm^2, 177.1 kg/hm^2 and 236.0 kg/hm^2, respectively, and the contribution rates were thus 39.3%, 42.4%, and 43.1%, respectively. Across these diffusion phases of soybean varieties, the contribution level increased gradually. The growth in the second period (1987 - 1994) was only 72.0 kg/hm^2, while it was 177.1 kg/hm^2 in the third phase, an increase of 146.0% over the previous phase, and 236.0 kg/hm^2 in the fourth period, an increase of 200 kg/hm^2 or 33.3% over the third phase. Since 2004, the first phase of variety upgrading has been observed to greatly improve the average yield of soybeans in China, with the related increase in yield of 329.4 kg/hm^2 exceeding 300 kg/hm^2 for the first time. This is also 39.6% higher than that seen in the previous stage and 4.6 times higher than that in the second diffusion phase. In terms of the variety contribution rate to yield increases, however, soybean showed a significantly different trend from those of rice, wheat, and corn. The contribution rate of yield increase showed a gradual increase from 39.3% in the second diffusion phase to 47.4% at the current diffusion period, increased by a large margin at 8.1 percentage points.

CHAPTER 6 The contribution of seed variety to cotton yield increases

6.1 Cotton yield variation in China

China's cotton production overall demonstrates a wavelike rise trend from 1980 to 2020. Cotton production in 1980 was 2.707 million tons, an increase of 3.203 million tons compared with 5.91 million tons in 2020, with an average annual growth rate of 2%. From 1980 to 2000, cotton production fluctuated significantly, the largest increases occurring in 1984 and 1991, when production reached 6.258 million tons and 5.675 million tons, respectively. From 2000 to 2020, cotton production increased overall, though the highest production of 7.232 million tons was reached in 2008. Since then, cotton production has stabilized to an overall upward trend.

Cotton yields showed rapid growth trends that fluctuated significantly over the period, while cotton yield fluctuated in line with the production from 1980 to 2000, with the largest increases in 1984 and 1991 for yields of 904 kg/hm^2 and 868 kg/hm^2, respectively. Cotton yield growth then accelerated, rising from 1,093 kg/hm^2 in 2000 to 1,865.15 kg/hm^2 in 2020, giving an increase of 70.6% in 20 years (Figure 6-1).

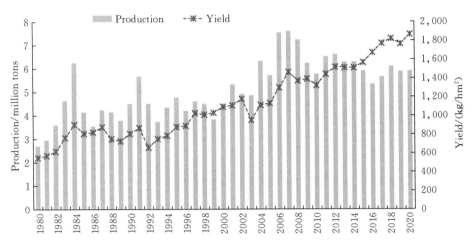

Figure 6-1　Changes in cotton yield and production, 1980-2020

Source: National Bureau of Statistics.

6.2　Evolution in cotton seed varieties in China

From 1980 to 2019, seed diffusion for cotton varieties experienced six periods. In the first period, ranging from 1980 to 1989, the leading variety was Lumian No. 1 (LM1). The second period lasted for 6 years, from 1990 to 1995, as Zhongmian No. 12 (ZM12) became the leading cotton variety nationwide, while the third period was from 1996 to 1998, during which Simian No. 3 (SM3) began to be most extensively diffused across China. The fourth period ranged from 1999 to 2003, when the leading variety changed to Xinmian No. 33 (XM33), and the fifth period was from 2004 to 2007, the major variety was changed to Lumian No. 21 (LM21); finally, the sixth period ranged from 2008 to 2019, when the top seed variety was replaced by Lumian No. 28 (LM28) (Figure 6-2).

CHAPTER 6 The contribution of seed variety to cotton yield increases

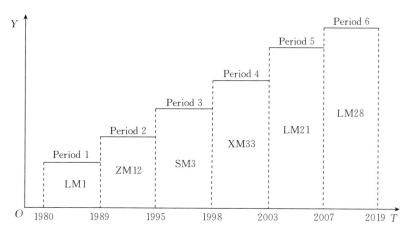

Figure 6-2 Leading varieties and diffusion period for cotton, 1980 – 2019

6.3 Specification for the cotton yield model

Due to the diffusion periods for seed varieties being divided into six distinct periods between 1980 and 2019, cotton yield can be specified as a model by including five dummy variables into the basic yield response equation:

$$Y_{it} = C + \beta_0 Y_{it-1} + \beta_1 P_{it-1} + \beta_2 D_2 + \beta_3 D_3 + \beta_4 D_4 + \beta_5 D_5 + \beta_6 D_6$$

(6-1)

where Y indicates yield; P represents producer prices; i indicates the main producing provinces in China; t is the year, from 1980 to 2019, such that $t-1$ represents the previous year, lagged; and D_2 to D_6 are dummy variables distinguishing different diffusion periods. D_2 denotes seed diffusion Period 2, from 1990 to 1995, by taking a value of 1, and otherwise taking a value of 0. Similarly, D_3 represents diffusion Period 3, from 1996 to 1998 as 1, otherwise, 0;

D_4 represents diffusion Period 4, from 1999 to 2003 as 1, otherwise, 0; D_5 represents diffusion Period 5, from 2004 to 2007 as 1, otherwise, 0; and D_6 represents diffusion Period 6, ranged from 2008 to 2019 as 1, otherwise, 0.

6.4 Model estimation

6.4.1 Estimation data

The estimation uses provincial panel data for cotton from 1980 to 2019 in China. Data on cotton yields are obtained from the China Rural Statistical Yearbook. Price data are drawn from the National Cost and Return of Agricultural Products in China and are expressed in real terms with 1980 as the base year. Information on the major varieties diffused in each province is from the annual internal reports from 1980 to 2019 provided by the Seed Administration of the Ministry of Agriculture and Rural Affairs of China (Table 6-1).

Table 6-1 Descriptive statistics for cotton yield and prices

Variable	Mean	Std. Deviation	Minimum	Maximum
Yield	968.37	391.77	157.50	2,340.63
Price	148.92	30.69	94.84	252.83

6.4.2 Estimation results

As the yield equation is a dynamic panel model, the Arellano-Bond method is used to estimate the equation to correct for regression bias caused by endogeneity. The estimated results are reported in Table

CHAPTER 6 The contribution of seed variety to cotton yield increases

6-2. The dummy variable for Period 2 is significant at the 10% level, while the variable indicating Period 3 is significant at the 5% level, and the variables for Periods 4 to 6 are highly significant at the 1% level. All variables take positive signs, suggesting that various varieties have contributed to cotton yield increases. The parameters on the dummy variables represent the average increases in cotton yields that may be attributed to the diffusion of the new variety over the average yield in the base period.

Table 6-2 **Estimated results of the yield response model**

Variable	Coefficients	Std. Error	Prob.
Cons.	223.09	65.04	0.001***
Y_{t-1}	0.61	0.05	0.000***
P_{t-1}	0.37	0.27	0.163
D_2	35.75	19.33	0.064*
D_3	68.54	28.14	0.015**
D_4	117.81	24.41	0.000***
D_5	185.64	31.19	0.000***
D_6	215.00	34.93	0.000***
Obs.	786	Wald	699.16

Note: ***, **, and * indicate the 1%, 5%, and 10% significance level, respectively.

The parameter for D_2 is 35.75 which implies that new seed diffusion in Period 2 (1990 – 1995) increased the average yield by 35.75 kg/hm² over the average in the base period (1980 – 1989). Likewise, in Period 3 (1996 – 1998), the average increase was

68. 54 kg/hm², a net increase of 32. 79 kg/hm² or 91. 7% over the increase in Period 2, indicating that cotton yields in this period were promoted by 68. 54 kg/hm² relative to the base period. In Period 4 (1999 – 2003), the rise was 117. 81 kg/hm², an increase of 49. 27 kg/hm² or 71. 9% over that in the previous period. In Period 5 (2004 – 2007), the increase due to new seed diffusion jumped to a high of 185. 64 kg/hm², an increase of 67. 83 kg/hm² or 57. 6% over that in Period 4. Finally, in Period 6 (2008 – 2019), variety replacement contributed 215. 00 kg/hm² to cotton yield increase, a slight increase of 29. 36 kg/hm² or 15. 8% over the previous period. Overall, the average increase in cotton yields due to the diffusion of new varieties over the base period rose over time, indicating a gradual improvement in cotton seed breeding and diffusion.

6.5 Empirical results

6.5.1 Absolute contribution of cotton varieties

Absolute contribution is the net increase attributed to crop variety replacement based on a unit of kilograms per hectare. Figure 6-3 illustrates the absolute contributions of varieties to cotton yield increases. The base period for cotton variety diffusion was from 1980 to 1989 when the leading cotton variety was Lumian No. 1; in Period 2 (1990 – 1995), as Zhongmian No. 12 became the major cotton variety, the average cotton yield was increased by 34. 1 kg/hm²; in Period 3 (1996 – 1998), when Simian No. 3 was extensively diffused nationwide, the average cotton yield was increased by 72. 7 kg/hm², an increase of 38. 6 kg/hm² or 113. 2% over the contribution made

CHAPTER 6 The contribution of seed variety to cotton yield increases

by the variety in Period 2. In the fourth period (1999 – 2003), cotton yield saw a net increase of 143.3 kg/hm² over the yield in the base period due to the cotton variety replacement, a jump of 70.6 kg/hm² or 97.1% over the contribution in the third period, indicating that the cotton variety diffused in this period, Xinmian No.33, significantly promoted cotton yield levels. In the fifth period (2004 – 2007), the net increase in yield caused by variety replacement of Lumian No.21 was just 176.5 kg/hm², only 33.2 kg/hm² or 23.2% higher than the contribution of variety in the previous period, so that the increase in the absolute contribution was the lowest among all variety diffusion periods; however, in Period 6 (2008 – 2019), the contribution of variety to yield increase jumped to a high of 215 kg/hm², a net increase of 38.5 kg/hm² or 21.8% over the fifth diffusion period for cotton varieties, indicating a gradual progress in cotton breeding technologies and seed diffusion work in this period (Figure 6-3).

6.5.2 Contribution rate of cotton varieties

Table 6-3 reports the detailed changes in cotton yield and the contributions of new varieties to this. The yield average in the base period (1980 – 1989) was 673.2 kg/hm², and this increased to 776.4 kg/hm² in Period 2 (1990 – 1995), an improvement of 103.2 kg/hm²; of this, new varieties contributed 34.1 kg/hm² to the total increase, with the contribution of the new seed varieties thus equaling 33.0% in this diffusion period. In Period 3, the average yield of cotton grew to 883.5 kg/hm², an increase of 210.3 kg/hm² over the average in the base period, of which 72.7 kg/hm² was

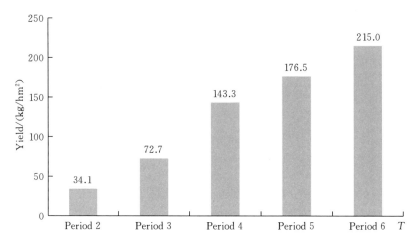

Figure 6-3　Absolute contribution of varieties to cotton yield in the base period

Note: cotton varieties contain six periods, which are 1980 - 1989, 1990 - 1995, 1996 - 1998, 1999 - 2003, 2004 - 2007, and 2008 - 2019, respectively.

attributed to new seed varieties; thus, the contribution of new seeds for the third diffusion period was 34.6%. In Period 4, the average cotton yield was 982.9 kg/hm², 309.7 kg/hm² higher than the base period, with new seed diffusion estimated as being responsible for 143.3 kg/hm² of the total increase, resulting in a contribution rate of 46.3% for this period. This was a period in which an increase was most clearly seen in the proportion of improvement attributed to new seed diffusion, with this contribution 11.7 percentage points higher than that in the previous period. There was then a significant decrease in the contribution of new seed varieties in the fifth diffusion period, with the increase due to new variety diffusion being just 33.2 kg/hm² or 23.2%, from 143.3 kg/hm² to 176.5 kg/hm². The total increase in cotton yield average was increased to 1,089.3 kg/hm², a jump of 416.1 kg/hm² from the base period, but the contribution rate decreased to 42.4%, 3.9 percentage points lower than that in the fourth period. In Period 6, the average yield for cotton grew to 1,252.2 kg/hm^2, an increase of 579.0 kg/hm^2 over the average in the base period, of which 215.0 kg/hm^2 was attributed to new seed varieties. Thus, the contribution of seed variety for the sixth diffusion period was 37.2%. These estimated results suggest that new seed varieties significantly contribute to rises in cotton yields, as well as implying that development in seed technologies and seed diffusion have succeeded in allowing significant progress in Chinese cotton production over the period under consideration.

Table 6-3　Varietal contribution to cotton yield increases over the base period (kg/hm^2, %)

Period	Yield	Yield increase	Varietal contribution	Contribution rate
Base (1980–1989)	673.2	—	—	—
2 (1990–1995)	776.4	103.2	34.1	33.1
3 (1996–1998)	883.5	210.3	72.7	34.7
4 (1999–2003)	982.9	309.7	143.3	46.4
5 (2004–2007)	1,089.3	416.1	176.5	42.4
6 (2008–2019)	1,252.2	579.0	215.0	37.2

Note: yield data are obtained from National Bureau of Statistics.

Table 6-4 displays the variety contributions as compared to the previous period. Therefone, the contribution rate for Period 2 is naturally the same as that over the base period, 33.1%. In Period 3, the average cotton yield was 883.5 kg/hm^2, 107.1 kg/hm^2 higher than the previous period. The increase caused by seed replacement over Period 2 was estimated as 38.6 kg/hm^2. Thus, the contribution of new seed varieties in Period 3 over the previous period was 36.1%. In the fourth period, cotton average yields exhibited a significant increase from 883.5 kg/hm^2 in the previous period to 982.9 kg/hm^2, a net increase of 99.4 kg/hm^2, while the

Table 6-4　Varietal contribution to cotton yield increases over the previous period (kg/hm^2, %)

Period	Yield	Yield increase	Varietal contribution	Contribution rate
Base (1980–1989)	673.2	—	—	—
2 (1990–1995)	776.4	103.2	34.1	33.1
3 (1996–1998)	883.5	107.1	38.6	36.1
4 (1999–2003)	982.9	99.4	70.6	70.9
5 (2004–2007)	1,089.3	106.4	33.2	31.2
6 (2008–2019)	1,252.2	162.9	38.5	23.6

Note: yield data are obtained from the National Bureau of Statistics.

CHAPTER 6 The contribution of seed variety to cotton yield increases

rise due to new seed diffusion grew to 70.6 kg/hm^2, causing the seed contribution rate to rise to 70.9%, 34.9 percentage points higher than that in the previous period. In the fifth diffusion period, a significant decrease was witnessed in seed contribution. However, the increase due to new seed diffusion fell from the previous 70.6 kg/hm^2 to 33.2 kg/hm^2, a decline of 37.4 kg/hm^2 or 52.9%. The cotton yield average in this period was 1,089.3 kg/hm^2, an increase of 106.4 kg/hm^2 or 10.8% over the average in period four. As a result, variety contribution in this period reached 31.2%, 39.8 percentage points less than that in the fourth diffusion period. Finally, seed variety's contribution reached a low of 23.6% in the sixth diffusion period, 7.6 percentage points less than that in the fifth diffusion period.

6.6 Summary

The examined diffusion period for cotton varieties can be split into six sub-periods, during which varieties of Lumian No.1, Zhongmian No.12, Simian No.3, Xinmian No.33, Lumian No.21, and Lumian No.28 were the leading varieties in turn. In the five diffusion periods after the initial period, the absolute contributions of the new varieties increased cotton yields per hectare by 34.1, 72.7, 143.3, 176.5, and 215 kg, and the contribution rates were 33.1%, 34.7%, 46.4%, 42.4%, and 37.2% for Period 2 to 6, respectively; the contribution rates of varieties thus sowed a generally increasing trend, indicating moderate improvement in production patterns for cotton. Concerning the contribution over the previous period, the absolute contributions were 34.1, 38.6, 70.6, 33.2, and 38.5 kg/hm², respectively, which were relatively stable, being in the range 30 to 40 kg/hm², except for the period four when Xinmian No. 33 was the most commonly diffused variety, though the respective contribution rates were 33.1%, 36.1%, 70.9%, 31.2%, and 23.6%, thus demonstrating a rising and then falling trend.

CHAPTER 7   The contribution of seed variety to rapeseed yield increases

7.1   Rapeseed yield changes in China

China's rapeseed yield and production levels have maintained growth for a substantial period. According to data from the National Bureau of Statistics, rapeseed yield in 1980 was 838.13 kg/hm^2, rising to 2,076.82 kg/hm^2 by 2020, an increase of 148%, which is in an average annual growth rate of 2.29%. From 1980 to 2004, the rapeseed yield fluctuated, increasing to 1,812.82 kg/hm^2 with an average annual growth rate of 3.27%; from 2004 to 2009, the yield steadily increased to 1,887.77 kg/hm^2 with an average annual growth rate of 0.81 %; in 2010, the yield decreased by 7.41% to 1,747.97 kg/hm^2; from 2010 to 2020, the yield steadily increased to 2,076.82 kg/hm^2 with an average annual growth rate of 1.74%. Rapeseed production in China increased from 2.38 million tons in 1980 to 14.49 million tons in 2020, which is an increase of 489%, or an average annual increase of 4.53%. The Chinese figures revealed cyclical fluctuations in rapeseed production during that time. Production increased by 6.17% annually from 1980 to 2009, to 13.54 million tons; in 2010, the negative growth rate was 5.52%, reducing production to 12.79 million tons; from 2010 to

2014, the average annual growth rate was 2.13%, which increased production to 13.91 million tons; in 2015 – 2016, the negative growth rate was 5.28%, so the production dropped to 13.13 million tons; from 2016 to 2020, the average annual growth rate was 1.71%, and production reached 14.05 million tons in 2020 (Figure 7-1).

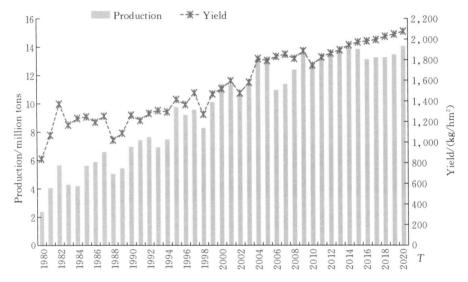

Figure 7-1   Changes in rapeseed yield and production, 1980 – 2020
Source: National Bureau of Statistics.

7.2   Evolution in rapeseed seed varieties in China

In terms of changes in major corn variety types during the period 1980 to 2019, the diffusion of rapeseed varieties can be apportioned into five distinct periods: the first period was from 1980 to 1986, when Xinan No. 302 (XN 302) was planted in vast areas; the second period was from 1987 to 2000, when Zhongyou No. 821

(ZY821) replaced XN 302 and become the leading rapeseed variety for more than 10 years; the third period ranged from 2001 to 2004, during this period ZY821 dropped from the top place, and Huaza No. 4 (HZ4) began to be diffused in most rapeseed producing areas in China, but this only lasted for a short period of 4 years; the fourth period was from 2005 to 2009, the major rapeseed variety became Qinyou No. 7 (QY7), which also only lasted for 4 years; and the fifth period was from 2010 to 2019, when the top rapeseed variety was changed to Qinyou No. 10 (QY10), where the diffusion period lasted 10 years (Figure 7-2).

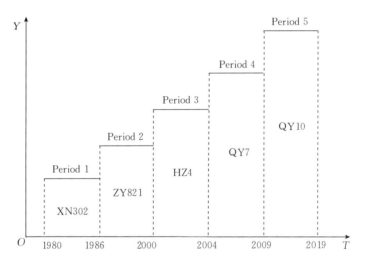

Figure 7-2 Leading varieties and diffusion period for rapeseed, 1980 – 2019

7.3 Specification for rapeseed yield model

The 5 distinct periods of changing rapeseed varieties between

1980 and 2019, allow rapeseed yield to be quantified as a model, by including four dummy variables into the basic yield response equation:

$$Y_{it} = C + \beta_0 Y_{it-1} + \beta_1 P_{it-1} + \beta_2 D_2 + \beta_3 D_3 + \beta_4 D_4 + \beta_5 D_5$$

(7-1)

where Y indicates rapeseed yield; P represents rapeseed producer prices; i indicates the main producing provinces in China; t is the year from 1980 to 2019 and $t-1$ represents the previous year, namely one-year lagged; and $D_2 - D_5$ are dummy variables distinguishing different diffusion periods. D_2 denotes seed diffusion period two, with 1987 to 2000 as 1, otherwise, 0; D_3 represents diffusion period three, with 2001 to 2004 as 1, otherwise, 0; D_4 represents diffusion period four, with 2005 to 2009 as 1, otherwise, 0; and D_5 represents diffusion period five, with 2010 to 2019 as 1, otherwise, 0.

7.4 Model estimation

7.4.1 Estimation data

Provincial panel data from 1980 to 2019 in China are used for estimation. Data on rapeseed yields are obtained from the China Rural Statistical Yearbook. Price data are drawn from the National Cost and Return of Agricultural Products in China and expressed in real terms, with 1980 as the base year. Information on major seed varieties diffused in each producing province is drawn from the annual reports published by the Seed Administration of the Ministry of Agriculture (Table 7-1).

CHAPTER 7　The contribution of seed variety to rapeseed yield increases

Table 7-1　Descriptive statistics for rapeseed yield and prices

Variable	Mean	Std. Deviation	Minimum	Maximum
Yield	1,535.37	516.83	225.00	2,932.54
Price	36.67	7.42	22.23	53.76

7.4.2　Estimation results

The Arellano-Bond method is used to estimate the coefficients, in order to correct the regression bias caused by endogeneity, and the estimated results are reported in Table 7-2. According to the estimated results, all variables are highly significant at the 1% level and are positive, strongly indicating that the model captures the net increases in rapeseed yield due to the changes in seed varieties. The parameters on the dummy variables represent the average increases in rapeseed yields which can be attributed to the diffusion of new varieties over the mean yield level in the base period.

Table 7-2　Estimated results of the yield response model

Variable	Coefficients	Std. Error	Prob.
Cons.	436.46	81.02	0.000***
Y_{t-1}	0.51	0.05	0.000***
P_{t-1}	2.91	0.62	0.000***
D_2	104.04	19.11	0.000***
D_3	298.22	32.02	0.000***
D_4	336.53	41.06	0.000***
D_5	416.14	40.85	0.000***
Obs.	588	Wald	961.16

Note: ***, **, and * indicate the 1%, 5%, and 10% significance level, respectively.

The parameter for D_2 is 104.04, which implies that seed variety dissemination in Period 2 (1987 – 2000) increased the average yield by 104.04 kg/hm² over the average in the base period (1980 – 1986). Similarly, in Period 3 (2001 – 2004), the average increase was 298.22 kg/hm², a surge of 194.18 kg/hm² or 186.6% over the increase in Period 2, indicating that rapeseed yields in this period were promoted by 298.22 kg/hm² relative to the base period. In period four (2005 – 2009), the rise was 336.53 kg/hm², an increase of 38.31 kg/hm² or 12.8% over that in the previous period. Finally, in period five (2010 – 2019), the proliferation due to new seed variety dissemination increased to a high of 416.14 kg/hm², a surge of 79.51 kg/hm² or 23.6% over the increment in the prevoius period. Over time, the average increase in rapeseed yields over the base period figures rose dramatically, due to the dissemination and usage of new varieties, which indicates a distinct improvement in seed breeding and distribution processes.

7.5 Empirical results

7.5.1 Absolute contribution for rapeseed varieties

The absolute contribution is the net increase attributed to crop variety replacement and with a unit of kilogram per hectare. Figure 7-3 illustrates the absolute contribution of varieties to rapeseed yield increases. The base period for rapeseed varieties diffusion is from 1980 to 1986, when the leading rapeseed variety was Xinan No. 302; in Period 2 (1987 – 2000), when Zhongyou No. 821 became the major rapeseed variety, the average rapeseed yield was

CHAPTER 7 The contribution of seed variety to rapeseed yield increases

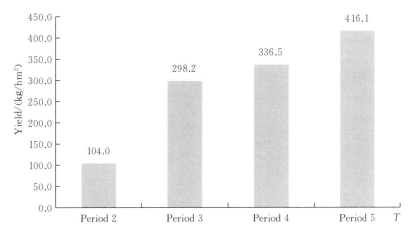

Figure 7-3 Absolute contribution of varieties to rapeseed yield in base period

Note: rapeseed varieties contain five periods, which are 1980 – 1986, 1987 – 2000, 2001 – 2004, 2005 – 2009, and 2010 – 2019, respectively.

increased by 104 kg/hm^2; in Period 3 (2001 – 2004), when Huaza No. 4 was extensively sown nationwide, rapeseed average yield was increased by 298.2 kg/hm^2, an increase of 194.2 kg/hm^2 or 186.7% over the absolute contribution made by the variety diffused in Period 2; in the fourth period (2005 – 2009), rapeseed yield was witnessed a net increase of 336.5 kg/hm^2 over the yield in the base period due to the rapeseed variety replacement, a rise of only 38.3 kg/hm^2 or 50.2% over the variety's contribution in the third period; in the fifth period (2010 – 2019), the net increase in yield caused by variety replacement was 416.1 kg/hm^2, 79.6 kg/hm^2 or 23.7% higher than the contribution of variety in the previous period, which indicates significant progress in rapeseed breeding technologies and seed distribution activity in this period.

7.5.2　Contribution rate of rapeseed varieties

Table 7-3 reports the detailed changes in rapeseed yield and the contribution of new varieties. The yield average in the base period (1980 – 1986) was 1,055.9 kg/hm^2, and it increased to 1,277.7 kg/hm^2 in Period 2 (1987 – 1994), an improvement of 221.8 kg/hm^2; new varieties contributed 104 kg/hm^2 to this increase, which is 46.9% of the yield increase in this diffusion period. In Period 3, the average yield of rapeseed grew to 1,580.5 kg/hm^2, an increase of 524.6 kg/hm^2 over the average in the base period, of which 298.2 kg/hm^2 was attributed to the adoption of new seed varieties, being 56.9% of the total increase in this period, namely, the contribution of new seeds for the third diffusion period was 56.9%. Period 3 is the only time when an increase was evident in the proportion of improvement attributable to new seed diffusion, with the contribution being estimated as 10.0 percentage points higher than the previous period. In period four, the average rapeseed yield was 1,784.5 kg/hm^2, 728.5 kg/hm^2 higher than the base period, and new seed diffusion was estimated as being responsible for 336.5 kg/hm^2 of the total increase, resulting in a contribution rate of 46.2% for this period, which is 10.7 percentage points lower than the contribution rate in the previous period. This reduction appears to be caused by a larger increase that has been attributable to other factors, such as plenty of input usage, which overshadowed the contribution share from new seed varieties and led to a decrease in relative seed contribution in this period. In the fifth diffusion period, with the increase due to new seed diffusion jumping from 336.5 kg/hm^2 to

CHAPTER 7 The contribution of seed variety to rapeseed yield increases

416.1 kg/hm², a net increase of 79.6 kg/hm² or 23.7%. The total increase in the average rapeseed yield was increased to 1,995.0 kg/hm², 939.0 kg/hm² up from the base period, while the seed variety's contribution rate decreased to 44.3%, 1.9 percentage points lower than that in the fourth period. These estimated results suggest that new seed varieties contribute significantly to advances in rapeseed yields, and that development in seed technologies and seed diffusion succeeded in making remarkable progress in China.

Table 7-3 Varietal contribution to rapeseed yield increases over the base period (kg/hm², %)

Period	Yield	Yield increase	Varietal contribution	Contribution rate
Base (1980 – 1986)	1,055.9	—	—	—
2 (1987 – 2000)	1,277.7	221.8	104.0	46.9
3 (2001 – 2004)	1,580.5	524.6	298.2	56.9
4 (2005 – 2009)	1,784.5	728.5	336.5	46.2
5 (2010 – 2019)	1,995.0	939.0	416.1	44.3

Note: yield data are obtained from the National Bureau of Statistics.

The seed variety contribution over the previous period for Period 2 is, therefore, the same as that over the base period, 46.9%, because the comparison is made between the same two periods. In Period 3, the average rapeseed yield was 1,580.5 kg/hm²,

302.8 kg/hm² higher than the previous period, and the increase caused by seed replacement over Period 2 was estimated as 194.2 kg/hm². Thus the contribution of seed variety in Period 3 over the previous period was 64.1%, and the variety's contributory share exceeded the sum of all other factors driving rapeseed yield. In the fourth period, average rapeseed yield only exhibited a slight increase, from 1,580.5 kg/hm² in the previous period to 1,784.5 kg/hm², a net increase of only 204.0 kg/hm², and the increase due to new variety diffusion shrank to 38.3 kg/hm², causing the seed contribution rate to drop to 18.8%, which is 45.3 percentage points less than that in the previous period. In the fifth diffusion period, a significant increase in seed contribution occurred. The increase due to new seed diffusion jumped from 38.3 kg/hm² in period four to 79.6 kg/hm², an upswing of 41.3 kg/hm² or 107.8%. The average rapeseed yield in this period was 1,995.0 kg/hm², an increase of 210.5 kg/hm² or 11.8% over the average in Period 4. Consequently, the seed variety contribution rate in this period reached 37.8%, 19.0 percentage points more than that during the fourth diffusion period, indicating that newly diffused varieties enhanced rapeseed yield significantly (Table 7-4).

Table 7-4 Varietal contribution to rapeseed yield increases over the previous period (kg/hm², %)

Period	Yield	Yield increase	Varietal contribution	Contribution rate
Base (1980–1986)	1,055.9	—	—	—

CHAPTER 7 The contribution of seed variety to rapeseed yield increases

(续)

Period	Yield	Yield increase	Varietal contribution	Contribution rate
2 (1987 – 2000)	1,277.7	221.8	104.0	46.9
3 (2001 – 2004)	1,580.5	302.8	194.2	64.1
4 (2005 – 2009)	1,784.5	203.9	38.3	18.8
5 (2010 – 2019)	1,995.0	210.5	79.6	37.8

Note: yield data are obtained from National Bureau of Statistics.

7.6 Summary

Rapeseed varieties have experienced five distinct diffusion periods since the 1980s, the respective leading varieties were Xinan No. 302, Zhongyou No. 821, Huaza No. 4, Qinyou No. 7, and Qinyou No. 10 for the five variety diffusion periods. The absolute contribution of the varieties to yield increases over the base period for Period 2, 3, 4, and 5 were 104.0, 298.2, 336.5, and 416.1 kg/hm^2, this last figure showing nearly a four-fold increase; the contribution rate of seed varieties to yield increase were 46.9%, 56.9%, 46.2%, and 44.3%, respectively. The contribution rate for rapeseed varieties is relatively stable, it generally remained at 45% for Period 2, 4 and 5, without significant deviation. For the contribution over the previous period, the absolute contribution of seed varieties used in Period 2,3,4 and 5 were 104.0,194.2,38.3, and 79.6 kg/hm^2, respectively; and the respective contribution rates for new seed varieties were 46.9%, 64.1%, 18.8%, and 37.8%, and the contribution rates over the previous period generally show a decreasing trend.

CHAPTER 8　Conclusion and policy recommendations

Assessing the contribution of new seed varieties to nationwide yield increases is a challenging but essential task. This study develops a method to estimate the contribution of new seed varieties to crop yield increases by constructing a yield response model which includs a series of dummy variables to capture the contributions of various varieties during their diffusion periods, as defined by the replacement of the leading varieties. Using the proposed method, the contribution of different varieties to yield increases for rice, wheat, corn, soybean, cotton, and rapeseed are estimated, and relevant policy implications are presented.

8.1　Conclusion

8.1.1　Absolute contributions keep increasing

The absolute contribution of varieties to crop yield increases refers to the yield increases caused by the adoption of new varieties, reflecting both the qualities of the genes bred into these new varieties and the extent to which their use is adopted around the country. The empirical results indicate that the absolute contribution of varieties to rice yield increased from 562.3 kg/hm^2 in the mid-1980s to 1,165.3 kg/hm^2

in the current period, representing an increase of 107.2%. The absolute contribution for wheat varieties increased from 235.7 kg/hm^2 in the early 1990s to 486.7 kg/hm^2, an increase of 106.5%. The absolute contribution of corn varieties increased from 406.9 kg/hm^2 in the late 1980s to 1,026.8 kg/hm^2 in the early 1990s, an increase of 152.3%. The contribution of soybean varieties increased from 72.0 kg/hm^2 to 329.4 kg/hm^2, an increase of 360%. The absolute contribution of cotton varieties increased from 34.1 kg/hm^2 to 215.0 kg/hm^2, an increase of more than 600%. The absolute contribution of rapeseed variety increased from 104.0 kg/hm^2 to 416.1 kg/hm^2, an increase of more than 400%. These results suggest that the absolute contribution of the six main grain varieties in China has maintained a stable growth trend since the early 1980s. In addition, the breeding and diffusion of varieties play a vital supporting role in increasing yield levels and ensuring the nation's food security.

8.1.2 Wheat varieties' contribution is relatively lower

The contribution rate of new varieties to crop yield increases is the contribution share of new varieties in total yield increases in a certain period, and is an important indicator measuring the progress and status of breeding technologies. The current contribution rates of new varieties to yield increases for rice, wheat, corn, soybean, cotton and rapeseed are 48.1%, 33.8%, 48.6%, 47.4%, 37.2% and 44.3%, respectively. Meanwhile, the contribution rates for rice, corn, and soybean are high and exceed 45%, while the contribution rate for wheat varieties is relatively low at under 35%. Among China's six main crops, wheat varieties contribute the least share to yield increases. As wheat is a vital staple food grain in China; thus, the low contribution rate of wheat varieties should arouse the attention of the Chinese government.

There are a number of factors that explain the low contribution rate of wheat varieties. Firstly, most wheat varieties are conventional varieties, and farmers generally use their wheat product as seed for the next season's production. This means wheat varieties can take a long time to update, possibly resulting in obvious degradation of the yielding potentials for wheat varieties. Secondly, great deals of resources are going towards rice and maize breeding programs, while wheat varieties have received relatively little attention. Inadequate investment restricts enthusiasm for breeding innovative wheat varieties. Thirdly, wheat production is not very profitable, so farmers are generally unwilling to adopt new varieties. This further reduces the contribution of modern varieties to wheat yield increases. The Chinese government needs to increase support for wheat variety breeding programs while improving wheat yield levels.

8.1.3 Contribution rates of main grain crops are in decline

The contribution rates of main grain varieties are on a long-term downward trend. In the early period, varieties made a strong contribution to yield increases; during the late 1980s and mid-1990s, rice's contribution rate once exceeded 70%, while corn approached 60%. The contribution rate subsequently experienced a downward trend, indicating that new varieties were becoming less important to promoting grain yield levels. At the current diffusion period, the contribution rate for rice varieties is 48.1%, which is 25.3 percentage points lower than the highest recorded level. Meanwhile, the current contribution rate for corn varieties is 48.6%, 10.5 percentage points lower than the historically highest level.

Crop varieties continue to contribute to yield increases. This means either that the absolute contribution of crop varieties keeps increasing, or that the numerator in the calculation formula of the contribution rate is increasing. However, the contribution rates appear to be decreasing. The only apparent reason for the decline in the contribution rates for crop varieties is that many input factors led to significant yield increases. The quantities of chemical fertilizers, pesticides, and other materials used today significantly exceed the levels used in the 1980s and 1990s, significantly enhancing crop yield levels. The contribution rate of varieties is a relative indicator. This means that when a certain factor input leads to a significant increase in yield (denominator), the contribution rate for crop varieties will decline even if the absolute contribution to yield growth of varieties continues to increase (numerator). It is necessary to strengthen the breeding programs of new environmentally-friendly varieties, especially fertilizer-saving and disease-resistant varieties. This means fewer chemical inputs will be needed in the crop-production process while also optimizing grain production and steadily increasing varieties' contribution rate, making agricultural production more environmentally-friendly.

8.1.4 Replacement in leading varieties is slow

The speed with which national leading varieties are replaced

CHAPTER 8 Conclusion and policy recommendations

will largely determine the contribution that new varieties make to yield increases. However, the replacement process of national leading varieties in China is relatively slow. In the case of wheat varieties, Bainong No. 3217 became the dominant variety in 1980 and was replaced by Yangmai No. 5 until 1989, which has been popularized for 10 years in China; the diffusion time for a wheat variety of Yangmai No. 18 reached 8 years, while the diffusion period of Jimai No. 22, the current dominant variety, exceeded 10 years. The slow upgrade of dominant wheat varieties resulted in the relatively low contribution rate of wheat varieties. One of the most important reasons for the slow replacement of dominant varieties may be the prominent phenomenon of variety self-keeping in grain production.

The survey found that the phenomenon of using own products of conventional rice and wheat as seeds for next production is very common in China. This may result in the slow replacement of the dominant varieties, and reduce the contribution rate of new varieties. To speed up the replacement of dominant varieties, the pace of scientific and technological innovation in the seed breeding areas needs to be accelerated. This will allow the creation of major new varieties, significantly improving China's food security situation. Further investment is also needed to improve the variety diffusion system.

8.1.5 Seed diffusion systems need to be improved

Agricultural extensions, including the diffusion of crop varieties in rural areas, have long been weak links in the chain of technological innovation. With the gradual liberalization of the seed market, the official agricultural agency's diffusion function has gradually weakened, and the original public diffusion system covering all rural areas has gradually disintegrated. Underfunding and understaffing are chronic problems in rural Chinese seed diffusion work, hindering the spread of new crop varieties. Due to the full liberalization of the seed market and lacking of an effective management system, there are massive varieties in the seed market. Commercial seed companies try their best to occupy the business share in the market competition, even launching price wars or lying about varieties' yielding capacities. As a result, a number of very ordinary seed varieties flooded the market, causing disruption to the normal order of the seed market. Farmers mostly focus on the price when selecting seed varieties, so ordinary varieties often beat the better varieties by virtue of being cheaper. The good varieties often struggle to be recognized by the market, resulting in "bad varieties drive out the good ones" and causing "market failure". Therefore, it is thus particularly urgent to construct a new diffusion system for good crop varieties, and strengthen the support to the public diffusion agencies.

8.1.6　Crop yield remains space to increase

Although China's crop yield levels have improved significantly since the 1980s, there is still significant room for further improvement. According to the latest statistics of the Food and Agriculture Organization (FAO), China's rice yield level was 7,056.2 kg/hm^2 in 2019, which was 84.3% of the United States (US), and also lower than Spain, Greece, and other coastal Mediterranean countries. China's wheat yield was 5,629.4 kg/hm^2 in 2019, which was only 63% of the United Kingdom's (UK) yield, and 72.7% of France's yield. China's corn yield was 6,317.1 kg/hm^2 in 2019, only accounting for 60% of the yield in the us, and also far behind Israel and other Middle East countries. China's soybean yield was 1,866.6 kg/hm^2, representing only 56% of Argentina's yield and 58.5% of the us's yield. There is a long way to go to narrow the yield gap between China and the world's highest levels. As China is restricted by its natural resources and environmental conditions, it is difficult to significantly increase the yield level by just using more input materials. However, there is significant potential for increasing the yield level by breeding improved crop varieties. China needs to accelerate the process of scientific and technological innovation in the seed industry, overcome the key technical obstacles in breeding technologies, and fully exert the yielding potential of good varieties.

8.2 Policy recommendations

8.2.1 Accelerating scientific and technological innovation in seed breeding

The government should intensify the implementation of major independent innovation projects relating to seed breeding, and strengthen the support to basic research in breeding and innovation in breeding theories and methods, such as systematic breeding, synthetic biology, genome editing, gene site-specific integration, and intelligent breeding. This will give China an advantage in breeding technologies and enhance the country's capacity for independent innovation in seed-breeding. In addition, efforts should be made to promote the application of biological breeding technologies in crop breeding. Researchers should aim to shorten the breeding cycle, improve breeding efficiency, and accelerate the cultivation of major national or regional crop varieties with independent intellectual property rights. The government should aim to establish national and regional seed innovation alliances, conduct joint studies on breeding national or regional varieties, and promote in-depth cooperation among regional seed breeding agencies, so as to optimize the allocation of cross-regional breeding resources, and create a collaborative development program for seed breeding science and technology between regions.

8.2.2　Reconstructing the promotion system of improved varieties

It is recommended to speed up the construction of a modern diffusion system for crop varieties, shoring up variety extension services in the innovation linkage of science and technology development. Commercial diffusion agencies will be the main body, with public diffusion agencies acting as an important supplement. This will allow for the rapid adoption of advanced crop varieties in agricultural production. We also recommend creating new diffusion methods and new mechanisms for variety diffusion, giving full play to the main role of seed enterprises in spreading new varieties. The government should strengthen the function of public diffusion agencies which focus on the diffusion of wheat and common rice. We should focus on demonstrating the usefulness of new varieties in

CHAPTER 8 Conclusion and policy recommendations

rural areas, and promote the construction of standardized demonstration bases for new varieties at the county and township levels. The government should also encourage seed enterprises and wholesalers to establish demonstration households for new varieties at the village level, in order to construct a network for spreading new varieties across the whole county. This will also gradually improve the variety diffusion system at the county, township, and village levels.

8.2.3 Supporting the breeding projects for green varieties

Attempts should be made to boost the research and development of environmentally-friendly crop varieties. These could be disease-resistant while also requiring less fertilizer, reducing the need for chemicals such as fertilizers and pesticides while increasing crop yields. Advanced breeding programs could also result in crops that require less water. They could also promote the integrated innovation for variety breeding technologies, as well as the breeding and diffusion of environmentally-friendly varieties. Attempts should also be made to broaden the variety test channels and improve the approval standards for new varieties. Under the basic condition of ensuring a stable yield level, a priority channel should be opened for the approval of green crop varieties. We should aim to construct producing bases for environmentally-friendly crop varieties, and improve the standardization, scale, and digitization of production in the bases. This would significantly increase production capacity for green varieties, ensuring the supply of environmentally-friendly crop varieties in the seed market. In addition, the demonstration and adoption of environmentally-friendly varieties would be faster, and promote the sustainable development of Chinese agriculture.

8.2.4 Matching improved varieties and fine planting method

Given the need to match good varieties with fine planting methods, it is crucial to spread knowledge about proper planting methods and patterns for various good varieties in the process of variety diffusion. This will increase the yield potential of these varieties, further boosting crop production. Through skill training, on-site observations, and integrated demonstrations of variety matching technology, we can teach the characteristics and correct planting methods for various varieties. Raising awareness of good varieties and corresponding farming methods will help farmers understand scientific planting and field management techniques. This will promote the use of new high-yielding, disease-resistant, and water-saving crop varieties, allowing them to move to full production. It is also important to promote cooperation between the departments of agricultural machinery, agricultural technology, and plant protection at the county level. Employees should receive professional skills training, and enhance their abilities of extension services on varieties and corresponding technologies, promoting the combination of agricultural machinery and agronomy. In addition, it is necessary to push the construction of high-quality farmland, and to improve the quality of medium-and-lowproductivity land, ensuring good varieties can reach their full yielding potential by providing favorable land conditions.

CHAPTER 8　Conclusion and policy recommendations

8.3　Suggestions for further research

This study develops a method for estimating the contribution of new seed varieties to crop yield increases, and conducted estimations to different crops. However, the study focused on variety changes at a national-wide level, and estimated the contribution of varieties across the whole country. Crop production conditions are multiple in China, meaning regional studies are necessary. Future studies may aim to consider changes in local varieties and explore the contribution of varieties to yield increases at the provincial level, although collecting relevant data might be challenging. In addition, studies using field trial data relating to new varieties can uncover the yielding potential of new varieties.

This method only estimates the comprehensive contribution of seed varieties in a given period, but this should be sufficient for the government to make an overall judgment on the status of seed variety development and diffusion. In the future, additional studies may aim to investigate the contribution of specific varieties and to separate the effects of diffusion from the overall contribution.

REFERENCES

ARELLANO M, STEPHEN B, 1991. Some Tests of Specification for Panel Data: Monte Carlo Evidence and an Application to Employment Equations [J]. The Review of Economic Studies, 58 (2): 277-297.

ASKARI H, CUMMINGS J T, 1977. Estimating agricultural supply response with the Nerlove model: A survey [J]. International Economic Review, 18 (2): 257-292.

BELL M A, FISCHER R A, 1994. Using yield prediction models to assess yield gains: A case study for wheat [J]. Field Crops Research, 36: 161-166.

BELL M A, FISCHER R A, BYERLEE D, et al., 1995. Genetic and agronomic contributions to yield gains: A case study for wheat [J]. Field Crops Research, 44: 55-65.

BRENNAN J P, 1984. Measuring the contribution of new varieties to increasing wheat yields [J]. Review of Marketing & Agricultural Economics, 52: 1975-1995.

CARGNIN A, SOUZA M A D, FRONZA V, et al., 2009. Genetic and environmental contributions to increased wheat yield in Minas Gerais, Brazil [J]. Science Agricola, 3: 317-322.

DE PONTI T, RIJK B, VAN ITTERSUM M K, 2012. The crop yield gap between organic and conventional agriculture [J]. Agricultural systems, 108: 1-9.

DUVICK D N, 2005a. The contribution of breeding to yield advances in maize (Zea mays L.) [J]. Advance in Agronomy, 86: 83-145.

DUVICK D N, 2005b. Genetic progress in yield of United States maize (Zea mays L.) [J]. Maydica, 50: 193-202.

EVANS L, FISCHER R, 1999. Yield potential: Its definition, measurement, and significance [J]. Crop science, 39 (6): 1544-1551.

FEYERHERM A M, KEMP K E, PAULSEN G M, 1989. Genetic contribution to increased wheat yields in the USA between 1979 and 1984 [J]. Agronomy Journal, 81: 242-245.

FEYERHERM A M, PAULSEN G M, 1981. Development of a wheat yield prediction model [J]. Agronomy Journal, 73: 277-282.

FEYERHERM A M, PAULSEN G M, SEBAUGH J L, 1984. Contribution of genetic improvement to recent wheat yield increases in the USA [J]. Agronomy Journal, 76: 985-990.

FISCHER R, 1975. Yield potential in a dwarf spring wheat and the effect of shading 1 [J]. Crop Science, 15 (5): 607-613.

FRENCH B C, MATHEWS J L, 1971. A supply response model for perennial crops [J]. America Journal of Agricultural Economics, 53 (3): 478-490.

FROSTER K A, MWANANMO A, 1995. Estimation of dynamic maize supply response in Zambia [J]. Agricultural Economics, 12 (1): 99-107.

GAFAR J, 1987. The supply response for sugar cane in Trinidad and Tobago: Some preliminary results [J]. Applied Economics, 19 (9): 1221-1231.

GIUNTA F, MOTZO R, PRUNEDDU G, 2007. Trends since 1900 in the yield potential of Italian-bred durum wheat cultivars [J]. European Journal of Agronomy, 27 (1): 12-24.

HE M X, HE C F, 1995. Measuring technical change in agriculture: The Analytical Hierarchy Process method [J]. Journal of Inner Mongolia Normal University, 2: 6-11.

HU R F, HUANG J K, ROZELLE S, 2002. Genetic uniformity and its impacts on wheat yield in China. Scientia Agricultura Sinica, 35 (12):

1442-1449.

MA Z Y, WU Y C, 2000. Contribution of rice genetic improvement to yield increase in China [J]. Chinese Journal of Rice Science, 14: 112-114.

MARTIN K L, HODGEN P J, FREEMAN K W, et al., 2005. Plant-to-plant variability in corn production [J]. Agronomy Journal, 97: 1603-1611.

MIAO R Q, MADHU K, HUANG H X. 2016. Responsiveness of Crop Yield and Acreage to Prices and Climate [J]. American Journal of Agricultural Economics, 98 (1): 191-211.

MIRSCHEL W, WIELAND R, WENKEL K O, et al., 2014. Yieldstat: A spatial yield model for agricultural crops [J]. European Journal of Agronomy, 52: 33-46.

MUSHTAQ K, DAWSON P J, 2002, Acreage response in Pakistan: A co-integration approach [J]. Agricultural Economics, 27: 111-121.

NERLOVE M, 1956. Estimates of the elasticities of supply of selected agricultural commodities [J]. Journal of Farm Economics, 27: 111-121.

NERLOVE M, BACHMAN K L, 1960. The analysis of changes in the agricultural supply: Problems and approaches [J]. Journal of Farm Economics, 42 (3): 531-554.

NEUMANN K, VERBURG P H, STEHFEST E, et al., 2010. The yield gap of global grain production: A spatial analysis [J]. Agricultural Systems, 103 (5): 316-326.

PATRIGNANI A, 2014. Yield gap and production gap of rainfed winter wheat in the southern Great Plains [J]. Agronomy Journal, 106 (4): 1329-1339.

PETERSON E W F, JIN L, ITO S, 1991. An econometric analysis of rice consumption in the People's Republic of China [J]. Agricultural Economics, 6 (1): 67-78.

QIAN J, ITO S, ISODA H, et al., 2012. Yield response to price and high-quality seed subsidy policies in China [J]. Japanese Journal of Farm

Management, 50 (1): 118-123.

QIAN J, ITO S, MU Y, et al., 2013a. High meat price and increasing grain consumption in China [J]. Japanese Journal of Farm Management, 51 (3): 67-72.

QIAN J, ITO S, MU Y, et al., 2013b. Impact of agricultural subsidies on rice price in China: A cointegration analysis [J]. Journal of Rural Problems, 49 (1): 172-176.

QIAN J, ITO S, MU Y, et al., 2015. Simulations on the impact of subsidy policies on grain supply and demand in China [M]. Beijing: China Agriculture Press.

QIAN J, ITO S, ZHAO Z, et al., 2015. Impact of agricultural subsidy policies on grain prices in China [J]. Journal of the Faculty of Agriculture Kyushu University, 60 (1): 273-279.

ROBERT S P, DANIEL L R, 1998. Econometric models and economic forecasts [M]. 4th editon. Beijing: China Machine Press: 242-245.

RUSSELL W A, 1991. Genetic improvement of maize yields [J]. Advance in Agronomy, 46: 245-298.

SAATY T L, 1980. The Analytic Hierarchy Process [M]. New York: McGraw-Hill.

SI W, LI D Y. 2018. Does variety extension affect China's soybean yield? [J]. Agrotechinical Economics, 5: 4-14.

SMALE M, HARTELL J, HEISEY P W, et al., 1998. The contribution of genetic resources and diversity to wheat production in the Punjab of Pakistan [J]. American Journal of Agricultural Economics, 80 (3): 482-493.

SPECHT J E, HUME D J, KUMUDINI S V, 1999. Soybean yield potential: a genetic and physiological perspective [J]. Crop science, 39 (6): 1560-1570.

TIAN H, GUO S, CHEN L, et al., 2006. The contribution of new varieties to increasing rapeseed yield [J]. Seed, 25: 73-76.

TIAN Z, JING Q, DAI T, et al., 2011. Effects of genetic improvements on grain yield and agronomic traits of winter wheat in the Yangtze River Basin of China [J]. Field Crops Research, 124: 417-425.

TIMMER C P, ALDERMAN H, 1979. Estimating consumption parameters for food policy analysis [J]. American Journal of Agricultural Economics, 61 (5): 982-987.

TOKATLIDIS I S, 2017. Adaptation to density to optimize grain yield: Breeding implications [J]. Euphytica, 213: 92.

TRAXLER, GREG, JOSE F-Z, et al., 1995. Production risk and the evolution of varietal technology [J]. American Journal of Agricultural Economics, 77 (1): 1-7.

TRAXLER, GREG, DEREK B, 1993. A joint-product analysis of the adoption of modern cereal varieties in developing countries [J]. American Journal of Agricultural Economics, 75 (4): 981-89.

VITALE D, DJOURRA H, SIDIBE A, 2009. Estimating the supply response of cotton and cereal crops in smallholder production systems: Recent evidence from Mali [J]. Agricultural Economics, 519-533.

WANG D, ZHANG Z, BAI C, 2007. Size of Government, rule of law, and the development of services sector [J]. Economic Research, J6: 51-64.

WANG L, HU R F, HUANG J K, et al., 2001. Soybean genetic diversity and production in China [J]. Scientia Agricultura Sinica, 34 (6): 604-609.

WANG T, LU C, YU B, 2011. Production potential and yield gaps of summer maize in the Beijing-Tianjin-Hebei Region [J]. Journal of Geographical Sciences, 21 (4): 677-688.

WANG W, LIAO Z, 2015. Experiment comparison for new late rice varieties in Ningdu County in 2014 [J]. Seed World, 8: 16-19.

REFERENCES

WANG X Q, CHEN H S, 1998. Analysis on vegetables supply response [J]. Inquiry into Economic Problems, 10: 54-56.

WATERFIELD D, 1985. Disaggregating food consumption parameters [J]. Food Policy, 10 (4): 337-351.

WOOLDRIDGE J M, 2013. Introductory econometrics: A modern approach [M]. 5th ed. Cincinnati: South Western.

WU Y C, MA Z Y, WANG D Y, et al., 1998. Contribution of maize improvement to yield increment in China [J]. Acta Agronomica Sinica, 24: 595-599.

XU X, HE P, ZHAO S, et al., 2016. Quantification of yield gap and nutrient use efficiency of irrigated rice in China [J]. Field Crops Research, 186: 58-65.

ZERE T B, VAN HUYSSTEEN C W, HENSLEY M, 2005. Development of a simple empirical model for predicting maize yields in a semi-arid area [J]. South African Journal of Plant and Soil, 22: 22-27.

ZHAO Z J, ZHANG S M, 2005. A quantitative analysis on factors driving agricultural technical progress [J]. Issues in Agricultural Economy, 1: 70-75.

图书在版编目（CIP）数据

农作物品种增产贡献率测算研究／钱加荣著．—北京：中国农业出版社，2023.9
　ISBN 978-7-109-31119-0

　Ⅰ.①农… Ⅱ.①钱… Ⅲ.①作物－良种－增产－研究－中国　Ⅳ.①S329.2

中国国家版本馆 CIP 数据核字（2023）第 177350 号

中国农业出版社出版
地址：北京市朝阳区麦子店街 18 号楼
邮编：100125
责任编辑：潘洪洋　　文字编辑：邓琳琳
版式设计：书雅文化　　责任校对：吴丽婷
印刷：北京中兴印刷有限公司
版次：2023 年 9 月第 1 版
印次：2023 年 9 月北京第 1 次印刷
发行：新华书店北京发行所
开本：700mm×1000mm　1/16
印张：8
字数：100 千字
定价：68.00 元

版权所有·侵权必究
凡购买本社图书，如有印装质量问题，我社负责调换。
服务电话：010-59195115　010-59194918